Palgrave Studies in Democracy, Innovation, and Entrepreneurship for Growth

Series Editor
Elias G. Carayannis, The George Washington University, Washington, DC, USA

The central theme of this series is to explore why some areas grow and others stagnate, and to measure the effects and implications in a trans-disciplinary context that takes both historical evolution and geographical location into account. In other words, when, how and why does the nature and dynamics of a political regime inform and shape the drivers of growth and especially innovation and entrepreneurship? In this socio-economic and socio-technical context, how could we best achieve growth, financially and environmentally?

This series aims to address such issues as:

- How does technological advance occur, and what are the strategic processes and institutions involved?
- How are new businesses created? To what extent is intellectual property protected?
- Which cultural characteristics serve to promote or impede innovation? In what ways is wealth distributed or concentrated?

These are among the key questions framing policy and strategic decision-making at firm, industry, national, and regional levels.

A primary feature of the series is to consider the dynamics of innovation and entrepreneurship in the context of globalization, with particular respect to emerging markets, such as China, India, Russia, and Latin America. (For example, what are the implications of China's rapid transition from providing low-cost manufacturing and services to becoming an innovation powerhouse? How do the perspectives of history and geography explain this phenomenon?)

Contributions from researchers in a wide variety of fields will connect and relate the relationships and inter-dependencies among (1) Innovation, (2) Political Regime, and (3) Economic and Social Development. We will consider whether innovation is demonstrated differently across sectors (e.g., health, education, technology) and disciplines (e.g., social sciences, physical sciences), with an emphasis on discovering emerging patterns, factors, triggers, catalysts, and accelerators to innovation, and their impact on future research, practice, and policy.

This series will delve into what are the sustainable and sufficient growth mechanisms for the foreseeable future for developed, knowledge-based economies and societies (such as the EU and the US) in the context of multiple, concurrent and inter-connected "tipping-point" effects with short (MENA) as well as long (China, India) term effects from a geo-strategic, geo-economic, geo-political and geo-technological set of perspectives.

This conceptualization lies at the heart of the series, and offers to explore the correlation between democracy, innovation and growth.

Milton M. Herrera
Editor

Business Model Innovation for Energy Transition

A Path Forward Towards Sustainability

Editor
Milton M. Herrera ⓘD
Economic Sciences Research Centre
Militar University of New Granada
Bogota, Colombia

ISSN 2662-3641 ISSN 2662-365X (electronic)
Palgrave Studies in Democracy, Innovation, and Entrepreneurship for Growth
ISBN 978-3-031-34792-4 ISBN 978-3-031-34793-1 (eBook)
https://doi.org/10.1007/978-3-031-34793-1

To my mom (Omaira), sister (Diana), brother (Duvan), and girlfriend (Anny) with love.

PREFACE

This book seeks to cover the gap in the existing literature on the business model innovation for energy transition. The book shows a dialogue between sustainability, technological innovation systems (TIS), simulation, green technologies index, dynamics performance management (DPM), and energy transitions. It also depicts how managers and policymakers can understand from modelling the dynamics of the energy transitions. The book is based on the premise that modelling business and dynamics of technological innovation can contribute to build a framework for sustainability transitions in the electricity industry. We present different contributions that can support our understanding of complex problems associated to dissemination of clean technologies with some case studies, particularly in Latin America. The methods described in this book have been used in a wide range of innovation studies and business issues. System dynamics modelling is presented in this book to gain a better understanding about the dynamics of the energy transition through business model innovation. This work is included in a series on *"Palgrave Studies in Democracy, Innovation, and Entrepreneurship for Growth"*.

Nowadays, the studies on sustainability transitions play an essential role in overcoming the problems associated with the climate change and wicked problems. As there are only a few studies on business model innovation applied to assess the energy transitions, this book offers a perspective based on business models. Besides, the book provides a methodological approach for measuring innovation of the green technologies, particularly in the renewable energy alternatives. These methods allow to foster the understanding and learning of the transition processes towards the sustainability in the energy sector. Therefore, this book separates the traditional methods used to model business and offers a novel way to explore the diffusion of renewables.

The book presents a novel method to design a green technologies index which can help to measure the technological innovation produced for the renewable energy industry. This book also explains the feedback perspective and the system dynamics method of modelling for representing the technological innovation from a DPM approach. This methodological approach can help to improve comprehension of value creation processes by pinpointing and studying the interconnections between strategic resource allocation, performance drivers, and outcomes. In this line, the book suggests and discusses the adoption of different tools as a way to overcome the barriers to business model innovation. Therefore, the book provides opportunities to identify strategic levers to increase the performance of the electricity industry by the depiction of dynamics of business.

This book stems from the collaboration of several scholars whose create an interdisciplinary work in which to explore the sustainability, innovation, and business modelling. Therefore, the book seeks to combine academic expertise in innovation studies, business management, energy studies, and policy analysis.

The book is divided into five chapters. The first chapter describes a way how the business model innovation may be represented to identify the performance drivers in the energy transitions. This chapter combines the concepts associated with the technological innovation system (TIS) and dynamic business modelling for sustainability (DBMfS) to identify strategic levers. The second chapter illustrates a macro perspective of the innovation process based on a dynamic performance management approach. This chapter examines the impact of various institutional conditions (exogenous) on the innovation (patents) as generator of economic growth. This research aims to create a DPM model to illustrate the dynamic behaviour of economic performance bounded by factor endowments, Foreign Direct Investment (FDI), and institutional conditions, to assist researchers and policymakers in gaining a deeper understanding of the economic complexity system. The third chapter develops a measure related to the National Innovations Systems literature. This chapter shows how a complex concept, such as the green technologies innovation system, can be measured. Besides, it proposes a new index for green technology innovation systems to rank regions and countries. The fourth chapter proposes a simulating model for public and private sector as a tool to understand sustainability behaviour caused by developing a project, in this case, a solar energy project. Finally, the fifth chapter seeks to assess the goal imposed by the Less Pollutant Urban Transportation Fleet Act (Law No. 16,802, of January 17, 2018), analysing the interaction between biodiesel, electric, hybrid, biomethane, and natural gas technologies to achieve reductions in CO_2, PM, and NOx.

Bogota, Colombia   Milton M. Herrera

PRAISE FOR *BUSINESS MODEL INNOVATION FOR ENERGY TRANSITION*

"This book provides a novel perspective and modelling-based solutions for energy transitions. The book is an essential read for decision makers and policymakers which seek sustainability transition."

—Mauricio Becerra-Fernandez, *Director of the Master in Smart and Sustainable Cities, Universidad del Rosario, Bogotá, Colombia*

CONTENTS

Notes on Contributors

Javier Andres Calderon-Tellez is a Ph.D. in Technology and Innovation Management (SPRU—Science Policy Research Unit) at University of Sussex Business School. He holds B.Sc. Mechatronics Engineering at the Nueva Granada Military University, M.Sc. degree in Project Management at the University of Sussex, and M.Sc. degree in Mechanical Engineering at the University of Massachusetts. He is now a Major in the Colombian Army. He has a scientific interest on system dynamics methodology, project management, and energy.

Ricardo E. Buitrago R. is a Research Professor in Strategy and International Management at EGADE Business School of Tecnológico de Monterrey. Furthermore, Dr. Buitrago is affiliated with the School of Management at Universidad del Rosario in Colombia as an associate researcher. He is a member of the Business Association for Latin American Studies (BALAS) Executive Committee. He is also a member of the Academy of International Business (AIB) and the Latin American Studies Association (LASA). He is also the ambassador for the Andean region of the IE-Scholars network. His research interests are focused on the junction between international political economy, institutions, strategy, and international business. He has served as editor and reviewer for the Journal of Business Research, International Business Review, European Management Journal, and International Studies of Management & Organization, among others.

Milton M. Herrera is currently Professor of Business Management at the Faculty of Economic Sciences, Universidad Militar Nueva Granada, where is also Senior Research Fellow in System Dynamics Modelling and the Director of Contemporary Studies Group in Organization Management at the Economic Sciences Research Centre. Professor Herrera is member of the System Dynamics Society.

He is author of the book series in *"Understanding Complex Systems"*, *"Lecture Notes in Energy"*, and *"System Dynamics for Performance Management"*, by Springer. He has published on several academic and professional journals, such as *Energy Journal, Renewable Energy, Utilities Policy, and Electricity Journal* among others, where and he also serves as a member of the Scientific Committee. He has also research experience with different research centres in public/private institutions for designing and implementing of policies and managerial strategies. He has a scientific interest on energy and food transitions, business model innovation, technological innovation systems (TIS), and supply chain performance.

Professor Herrera has been collaborating with several universities all over the world, ranging from Europe (University of Sussex and University of Palermo), and America (University of Santa Catarina).

Alberto Méndez-Morales is economist researcher and consultant in topics related to innovation economics, science and technology policies, innovation finance, intangible valuation, and patent quality. Broad experience in analysing science, technology, and innovation projects, consulting in valuing technological companies and intangible assets—lecturer in financial topics at internationally recognized universities. Professor Mendez is a researcher for EGADE Business School in México City.

Caroline Rodrigues Vas is an Associate Professor in the Department of Industrial and Systems Engineering at the Federal University of Santa Catarina, and a permanent faculty member of the Graduate Programme in Production Engineering at UFSC. She holds a Postdoc (PNPD/CAPES) in Production Engineering, Operations Management (2016–2018), and a Ph.D. in Production Engineering (CNPq) from UFSC (2016). She completed a research stay (PDSE/CAPES) at the Ecole Nationale d'Ingenieurs de Tarbes (ENIT) in France (2013). She also holds a Master's degree in Production Engineering (2010), an M.B.A. in Scientific and Technological Education (2008), an M.B.A. in Industrial Management: Production and Maintenance (2009), and a bachelor's degree in Food

Technology (2007), all from the Federal Technological University of Parana. She is a senior researcher at the Sustainable Energy Innovation Group (SINERGIA/UFSC). Her research areas include sustainability, renewable energy, and electric vehicles. She has published in journals such as Renewable and Sustainable Energy Reviews.

Mauricio Uriona Maldonado is an Associate Professor at UFSC in the Department of Industrial and Systems Engineering, in Florianopolis, Brazil. His research interests are in Renewable Energy, Electric Vehicles, Sustainability Transitions, and System Dynamics modelling. For the period 2022–2025 he is affiliated as Extra-Ordinary Associate Professor at the Department of Industrial Engineering, Stellenbosch University, South Africa. He is a senior researcher at the Sustainable Energy Innovation Group (SINERGIA/UFSC). He has been a visiting scholar at Universitat Bremen (Germany), University of Tampere (Finland), Universidad Autonoma de Madrid (Spain). He has received training in system dynamics at the Massachusetts Institute of Technology (USA) and in innovation economics at the University of Tampere (Finland). His previous project experience spans the Energy, Software, and Telecom sectors and as innovation management consultant at the Santa Catarina Industry Federation (FIESC). He holds a Ph.D. and an M.Sc. in Knowledge Management and Engineering at UFSC with co-supervision at Duke University (USA).

Tainara Volan holds a degree in Production Engineering from the Integrated Regional University of Alto Uruguai e Missões. She has a master's degree in Production and Systems Engineering from the Federal University of Santa Catarina. She is currently a Ph.D. Candidate, also in the Production and Systems Engineering programme at the Federal University of Santa Catarina. It employs technical forecasting methods (eg. system dynamics) to study socio-technical transitions, electric mobility, battery storage systems, and clean urban public transportation.

LIST OF FIGURES

LIST OF TABLES

Dynamic Business Modelling for Sustainability Transitions in the Electricity Industry

Milton M. Herrera ⓘ

Abstract Business models for socio-technical transitions are gaining increasing attraction by corporate sustainability scholars and practitioners who are currently discussing innovative methods and tools for supporting their design and use. The electricity industry has always played a crucial role in terms of sustainability impact, i.e. in the perspective of contributing to the social, economic, and environmental value development of a territorial area. The recent literature on business models design highlights a paucity of design tools and techniques oriented to consider the specific organizational and market characteristics of these firms. This paper shows how to apply a System Dynamics (SD) view in the design of a business model for the energy industry. The methodological support offered by SD modelling may improve the decision-making processes

M. M. Herrera (✉)
Economic Sciences Research Centre, Universidad Militar Nueva Granada, Bogotá, Colombia
e-mail: milton.herrera@unimilitar.edu.co

© The Author(s), under exclusive license to Springer Nature Switzerland AG 2023
M. M. Herrera (ed.), *Business Model Innovation for Energy Transition*, Palgrave Studies in Democracy, Innovation, and Entrepreneurship for Growth, https://doi.org/10.1007/978-3-031-34793-1_1

1

and the long-term impact associated with sustainable value creation using simulation scenarios and key performance indicators.

Keywords Business Model · System Dynamics · Electricity Industry · Sustainability · Simulation · Electricity Market · Renewables

1.1   INTRODUCTION

The energy transition involves facing new challenges to towards a more sustainable production based on renewable sources. This transition comprises large-scale deployment and effective integration of renewables, which could fundamentally change the structure of the power generation business and renovate the way how electricity is produced, transmitted, and sold (Castaneda et al., 2017a; Richter, 2013a; Zapata et al., 2023). Therefore, the business model innovation in all actors of the electricity industry is considered as a fundamental part of the solution to disseminate clean technologies both in developing and developed countries (Karami & Madlener, 2021; Trotter & Brophy, 2022; Visintainer et al., 2021). However, a systemic assessment of how renewable energy industry structure and operate their businesses in such countries is insufficient from the extant research.

Non-linear, disruptive systemic shifts that result from complex processes characterize the energy transitions (Ahlborg, 2017). Indeed, the complex nature of energy systems complicates the understanding of business contributions to sustainability transitions. In this sense, sustainability transition discussions enable to enrich the business model (BM) studies (Bocken & Short, 2016; Petzer et al., 2020; Sarasini & Linder, 2018). These discussions often focus on actors' roles and strategies in socio-technical transitions (Farla et al., 2012; Ruggiero et al., 2021). However, the adaptation to the modern energy schemes inspired by the change in consumers' preferences has created other business opportunities and jeopardized traditional models (Blazquez et al., 2020). For instance, the business model innovation might bring the change from electricity generation to create additional service in energy storage (Hamelink & Opdenakker, 2019).

The traditional business models of the electricity industry are facing emerging challenges around the world for several reasons. First, the traditional business model faces financial barriers both electricity firms and consumers associated to the lack of access to finance, high rates, and short payback periods (Eleftheriadis & Anagnostopoulou, 2015; Huijben & Verbong, 2013; Ruggiero et al., 2015). Second, the regulatory and institutional barriers that influence the power supply solutions, such as distributed energy deployment (Castaneda et al., 2017b; Horváth & Szabó, 2018; Jimenez et al., 2016). Third, the technological barriers associated to power grid availability that affect the security of supply (Herrera et al., 2019; Jacobsson & Karltorp, 2013; Richter, 2013a; Zapata et al., 2023). In this line, the electricity industry demands flexible BMs to adapt to the fluctuating market and socio-technical changes. In this context, the following questions may be relevant: How does the business modelling capture the feedback loops and the dynamic behaviour of vital performance measures? How are the electric utility industry prepared to follow up on their innovations and implement quicker sustainability concepts?

Within this context, as BMs explain the relationships among actors in a changing setting, the research on energy transitions assisted by a dynamic modelling approach can help towards truly sustainable BMs (Hirt et al., 2020). There is a burgeoning body of literature, which includes several articles on the business model innovation (BMI) (Sarasini & Linder, 2018). Research on this subject has increased, becoming a crucial topic to enrich the sustainability transitions field. The most relevant studies on sustainable business models emphasize the need for flexible models to respond to changing environments (Cosenz et al., 2019). Sarasini and Linder (2018) highlight new lines of inquiry that can examine the dynamics of BMI in the sustainability transitions. Bocken and Short (2016) remark on the need to foster the moderating end-user consumption as a driver of business model innovation for sustainability. In this line, the aim of this chapter is presented how the system dynamics (SD) methodology approach can aid to identify performance drivers of business model innovation for sustainability in the energy transitions.

Business models for energy transition are gaining increasing attraction by corporate sustainability scholars and practitioners who are currently discussing innovative methods and tools for supporting their design and use. Indeed, one of the current challenges for the electricity industry is to transform their business models in tune with shifting societal and market conditions. This study develops a dynamic business modelling based on

SD methodology for sustainability in the renewable energy industry. Furthermore, this chapter shows how performance drivers play an essential role to determine the business model innovation. In this sense, this study contributes to extent the body of knowledge on modelling of business model from a dynamic perspective for energy transition in a sustainability framework.

This chapter is organized as follows. Section 1.2 reviews the existing literature on modelling for business model innovation reported in the energy sector. Section 1.3 introduces the research framework and the methodology approach based on system dynamics simulation used in our study. Section 1.4 presents the results and discusses the main findings of our research based on case study. Finally, conclusions and practical implications are presented in Sect. 1.5.

1.2   RESEARCH ON THE BUSINESS MODEL INNOVATION FOR RENEWABLE ENERGY

At present, the electricity market is shifting from a model of pure electricity delivery to a market of mixed services and goods delivery, resulting in changes to the marketing activities (Sadjadi & Fernández, 2023). This shift to a more strategic sector is likely to impact the business models of firms involved in the energy storage market (Hamelink & Opdenakker, 2019). The business model innovation should consider emerging players as the energy market changes. Therefore, the electricity firms need to change traditional business models to innovative directions of the energy market.

Previous research on the business model innovation for renewable energy shows different trends associated with innovation and energy policy, sustainability, and investment, as illustrated in Fig. 1.1. This figure illustrates three clusters related to business model innovation and renewable energy. First, the studies on business models applied to energy sector, which involve aspects such as e-commerce and investment (Agrawal et al., 2022; Lozano & Reid, 2018). Secondly, energy policy considerations include studying business models for the dissemination of different energy sources (Ford & Hardy, 2020; Herrera et al., 2020a; Trotter & Brophy, 2022). Third, the sustainability issues aligned with renewable resources (Gsodam et al., 2015; Sadjadi & Fernández, 2023). Indeed, some of these studies reported the relationships between sustainability, innovation, and renewable energy (Calderon-Tellez et al., 2023). However, a few studies have not depth-in business model innovation for sustainability transition of electricity industry.

Most of the studies on business management agree that a business model can be useful for making decisions of a firm, creating sustainable competitive advantages in a specific sector (Castro, 2020; Ce, 2009; Cosenz, 2017; Cosenz & Bivona, 2020; Gronum et al., 2015). For instance, an innovation business model tested in China combined the use of solar photovoltaics and agricultural greenhouses unfilled a solution to meeting the potential for needed supply (Li & Shen, 2019). The business model also helps to expand the understanding of business phenomena and capture business value. In this line, Horváth and Szabó (2018) examined the development of photovoltaic business models to determine how the obstacles to distributed energy deployment can be addressed through the community-shared model. This study shows a comparison with other solution alternatives such as host owned, and third party owned. According to another study, an increased investment subsidy will

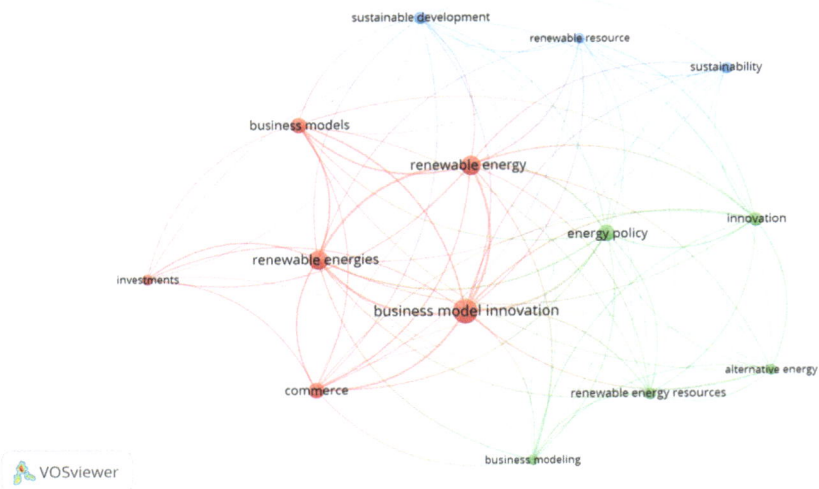

Fig. 1.1 Map of keywords based on co-occurrence with the assistance of VOSviewer

motivate the company to make sales, while a larger generation subsidy will make non-ownership models more appealing (Agrawal et al., 2022). In this line, as large investment will be needed to make the electricity sector more sustainable, the investors will have a key role in funding innovation and achieving such changes for the energy sector (Lozano & Reid, 2018).

It is essential to contemplate new ways of creating value for decarbonization of electricity industry through development of innovate business model. Richter (2013b) identified opportunities and barriers to business model innovation for the German utilities and distributed PV. The main conclusion of this study is that utilities could benefit from changing their perspective on distributed PV when seen as only just another source of electricity production. Another similar study developed by Bukari et al. (2021) concludes that the biggest obstacle to accelerating the deployment of mini-grids is severe lack of funding for electricity utilities, as it limits business model innovation. Horváth and Szabó (2018) highlight the importance of reviewing the strategic assets and key competences of utilities to implement an innovate business model. Besides, this study shows that the community-owned model can generate significant benefits in many areas, and trends such as increasing digitization.

The technological progress together with changing energy market is making it difficult for organizations to keep up with traditional ways of doing business. Smart energy products and services can lead to more efficient use of energy resources. Chasin et al. (2020) provide an analysis of the smart energy market for private households. This study contributes as a starting point for business model innovation especially for electricity utilities. Currently, the renewable energy debate shows how the technological progress contributes to decentralized electricity generation based on renewable energies by smart grids, new storage technologies, energy efficiency solutions, and more active customers (Bellekom et al., 2016; Gsodam et al., 2015; Rehman et al., 2023; Weigelt et al., 2021).

Despite several studies on business model innovation have been developed, there are not enough studies that analyse the dynamics between actors—feedbacks or delays—for sustainability transition in the electricity sector. Additionally, there has not been a thorough examination of what key performance drivers are necessary to put the business model into practice.

1.3 Methodology Approach

Learning based on experimentation is an important capability in the transition to a sustainable business. The learning is a feedback process in which our decisions alter the actual world. The SD modelling has proven to be useful to enhance learning about social systems. While other traditional approaches have limited support to draw the innovation processes (Holtz et al., 2015), SD has been successful in experimenting with models in an innovation context (Herrera & Trujillo-Díaz, 2022). Besides, it provides a very schematic and formalized view of the regime concept and multi-level dynamics (Van Den Bergha, 2012).

In this context, this chapter developed a simulation model based on SD methodology approach. SD is particularly useful for gaining insight into an energy system characterized by delays, non-linearity, and feedbacks (Ford, 1997; Morcillo et al., 2017; Sterman, 2000, 2003). Moreover, this simulation approach allows the adoption of a systemic perspective for mapping value generation processes (Calderon-Tellez et al., 2023; Herrera & Trujillo-Díaz, 2022). For instance, Cosenz (2017) remarks that Business Model Canvas (BMC) may be associated with limitations in evaluating the sustainability of a business. Meanwhile, SD uses simulation scenarios to provide a better understanding of sustainability.

This section shows two steps used to develop a business model based on an SD methodology approach for the transition of the electricity industry that integrates the concepts of "motor innovation" and the dynamic business modelling for sustainability (DBMfS) framework. The motors of innovation contribute to understanding the interactions among actors and institutions (Hekkert & Negro, 2009; Hekkert et al., 2007; Markard et al., 2012; Suurs, 2009). This concept addresses technological change as being a complex non-linear interactive process for developing and diffusing innovations (Herrera & Trujillo-Díaz, 2022; Uriona & Grobbelaar, 2019). The main idea behind the "motors of innovation" is the design and analysis of the innovation system (IS). First, a stock and flow diagram was used to depict the four motors of an IS proposed by Suurs (2009): the motor of science and technology (R1–R2), the market motor (B1–B2), the entrepreneurship motor (R4), and the system building motor (R3), as presented in Fig. 1.2. The stocks represent the resources at a specific moment in time, while a flow is viewed as a rate of change over time (Forrester, 1997). In this line, Suurs (2009) argues that the motors of innovation represent phenomena into IS that are subject to change over time, which may be represented with the approach of SD.

The attraction of the product is one of the considerations for clean technology diffusion—product attractiveness (PA), which is stimulated by margin market (MM). This driver also influences on desired investment for the electricity transition. The driver is defined in this model below:

$$PA(MM) = \begin{cases} 0.1, if\, MM = 0 \\ 0.25, if\, 1 > MM \geq 0.5 \\ 1.2, if\, 1.6 \geq MM > 1.1 \end{cases} \qquad (1.1)$$

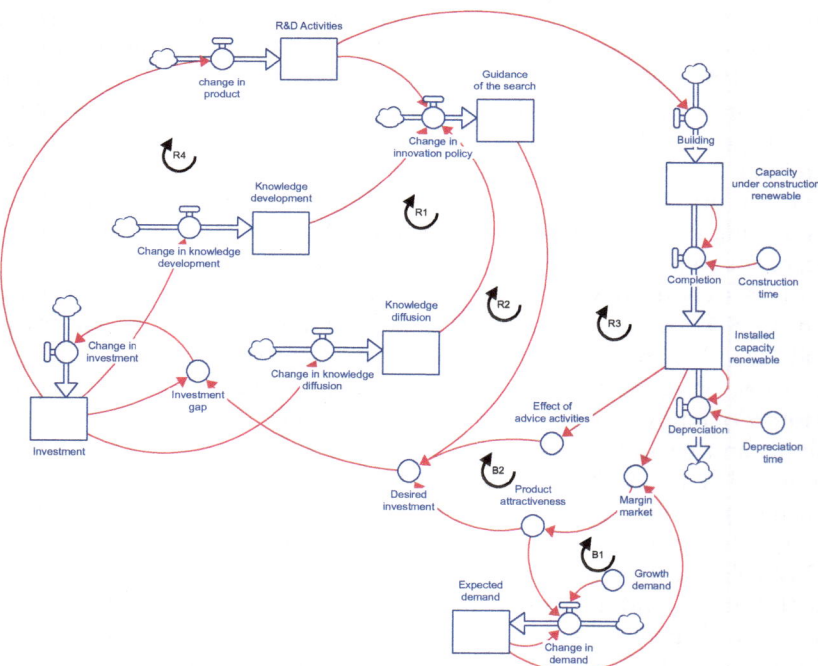

Fig. 1.2 Stylized overview of the SD model for assessing the sustainability transition and innovation system in the electricity sector

Second, Cosenz et al. (2019) proposed a dynamic framework for business modelling—named dynamic business modelling for sustainability (DBMfS). The dynamic framework has shown to be effective in mapping business systems characterized by dynamic complexity and unpredictability. This framework emphasizes the pathway more than the results. Besides, DBMfS might reach the challenge declared by McDowall and Geels (2017) due to the fact that it can establish a constructive dialogue between model-based scenarios and narrative scenarios by pluralist unified strategies. At this stage, the methodology identifies the system's performance drivers. In sum, the SD model allows the testing of the system's performance (first step), considering the identification of the performance drivers for the business (second step).

1.3.1 Case Study

The Brazilian energy sector suffered structural reforms in 2004, with the introduction of the auctions system for energy contracting (Dutra & Szklo, 2008; Mastropietro et al., 2014). As wind power auctions have fostered the participation of local suppliers in the electricity market, the investors' confidence of wind technology has increased. Brazil has experienced significant advances in its energy policy, and wind industry has shown a high growth rate from 14 GW capacity in 2018 (ABEEólica, 2018a; Pereira et al., 2011) to 25 GW in 2023 (ABEEólica, 2023). The promotion of auctions for wind power contracting is an overarching factor of the wind industry in Brazil. The Brazilian government and public companies related are responsible for promoting the regulated auctions guidelines, as well as determining the values auctioned. The contracting of wind power is regulated by energy auctions (beginning in 2009), which is critical for the development of this industry in the country.

In the case of Brazil, the Brazilian National Development Bank (BNDES) is the main financing entity for new wind farms, almost 90% of the existing wind farms were built from its financing (Bradshaw, 2017; Porrua et al., 2010). Still, it should be noted that since 2012, the percentage of investment in renewable energy sources was primarily dedicated to wind energy. For example, in 2012, investments made in the wind sector accounted for ~46% of the country's total investments in renewable energy, growing annually, until reaching a 58% share by the end of 2017. This steady increase in investment is essential for Brazil to garner attention from signal investors, giving them confidence to invest in the entire wind power chain (ABEEólica, 2018b).

1.4   RESULTS AND DISCUSSION

This section shows the results obtained from business modelling for the case study described above. Consequently, the DBMfS was designed based on a stock and flow diagram, which is shown in Fig. 1.3, and is used to simulate and represent the sustainability transition in the energy sector. Keeping the same variables identified in Sect. 1.3, the built-in stock and flow diagram is remodelled and simulated according to the case study described. This stage offers a qualitative understanding of the dynamic business for sustainability canvas in the energy sector.

From the DBMfS designed, it is possible to identify as stocks—the strategic resources and key processes represented (i.e. rectangle-shaped

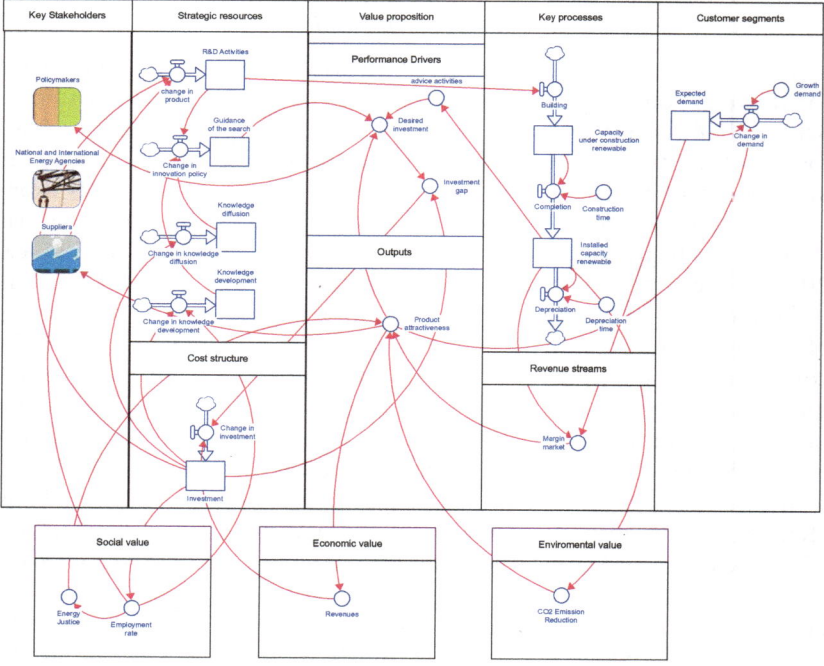

Fig. 1.3 The dynamic business modelling for sustainability canvas applied to data of case study

variables). Particularly in this case study, the strategic resources are associated with the functions of innovation system modelled by Walrave and Raven (2016). Additional to this, the key processes are represented by the changes of energy capacity of the system along the supply chain (Herrera et al., 2018). A strategy lever identified as desired investment it defines the strategic planning from resources and processes of the energy supply chain. In other words, the managers and policymakers may define the business strategy when modifying budget allocation or investment.

The strategic resources affect a set of drivers such as investment gap and desired investment. These drivers are modelled as circle-shaped variables in the DBMfS designed. Value proposition and outputs produce a set of results for economic, social, and environmental aspects. In the case study, the values related to sustainability transition in the energy sector may relate to employment rate, revenues, and CO_2 emissions reduction. Eventually, these values contribute to improve the product attractiveness, which leads to an increase in the installed capacity of renewable energy in a way sustainable in the long run. Expanding and identifying the value propositions could allow utilities to earn higher revenues per customer (Richter, 2013b). Besides, it is possible under certain conditions to attain a balance between social welfare and the barriers to business model innovation, which leads to overcome the death spiral of utilities (Castaneda et al., 2017b).

Figure 1.4 shows four simulation scenarios designed for assessing the installed capacity of renewables. The first scenario represents market changes where margin could reach between 1.1 and 1.6. When the capacity margin rises cause a decrease in the electricity price, which helps to increase the capacity of system. This scenario represents a large-scale dissemination of renewables. Meanwhile, the second scenario contemplates a capacity margin between 0.5 and 1. This scenario may occur if the generation capacity margin fall, then generates a high price might be associated, for example, with insufficient transmission capacity (Herrera et al., 2020b). The fourth scenario represents the current situation of the system. Finally, the fifth scenario considers a capacity margin equal to cero. This pessimist scenario is associated with a decrease in investment incentives to expand of renewables. In sum, the results show that the first scenario represents an optimistic behaviour in related with the increase of renewables capacity. In contrast with the fourth and fifth scenarios which present a pessimistic behaviour in the installed capacity of renewables.

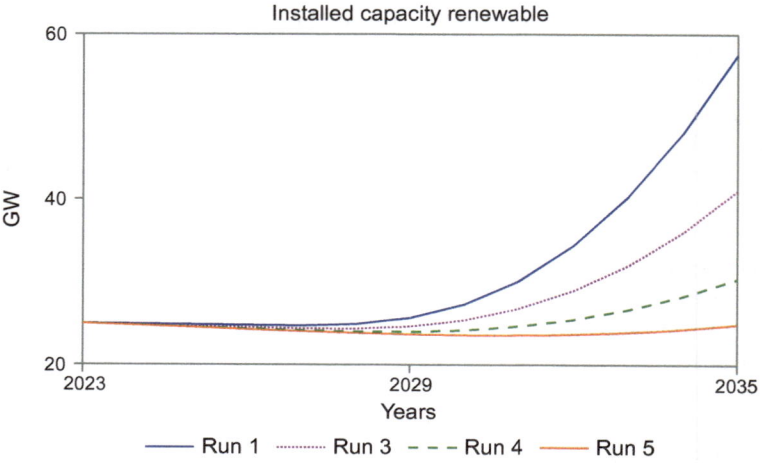

Fig. 1.4 Results obtained for the installed capacity of renewables

1.5   Conclusion

This chapter presented as SD modelling may improve the decision-making processes and the long-term impact associated with sustainable value creation using simulation scenarios and key performance indicators. The energy transitions have not been fast enough to avoid the impacts of climate change (Herrera et al., 2020c; Su et al., 2023; Xi et al., 2023). Several streams of future research should be explored in BMI; in other words, the development of a quantitative approach based on the conceptual core of the sustainability transitions field. In this way, this chapter argues that BMI for energy transitions research needs to comprise a dynamic perspective, which could be included in future discussions.

Findings show that the key performance indicators play an essential role in the electricity industry and have significant implications for theory and policy for sustainability. Since the global warming problem threatens the whole world, it is needed that countries should develop energy policies that will increase their sustainable and clean energy investments (Carayannis et al., 2022a). Policymakers are better able to implement effective measures to increase the share of renewable energy in the national electricity mix when they have identified the reasons and drivers for developing new business models for electricity generation from renewable sources (Gsodam et al., 2015). This implies that for a TIS to be a Quadruple and Quintuple Helix Innovation System as proposed by (Carayannis et al., 2022b), the government and the political system should be based on democracy and ecology. Indeed, it should also be regarded as drivers for further knowledge production and innovation development. This study also shows that identifying performance drivers in the business model discloses vital additional information on how the investment in the electricity sector influences on the increase of new installed capacity of the system. Investing in distributed energy projects could be made even more attractive by taking advantage of the knowledge economy to promote research and innovation (Carayannis et al., 2022a).

There are two limitations that should be considered when analysing the present case study results. First, this paper only represented a case study in Brazil with data of wind energy market. This study does not account for how the business model could be applied to assess strategies in other countries, where modelling in terms of the market conditions could raise a different set of issues. Second, the availability of data can cause concern for the validation of the simulation model proposed, even more so because some of these data are qualitative (especially in innovation functions) (Holtz et al., 2015).

Acknowledgements The authors are grateful to Universidad Militar Nueva Granada (Grants IMP-ECO-3402) for providing the necessary financial support for this chapter.

References

ABEEólica. (2018a). Dados ABEEólica.

ABEEólica. (2018b). Annual wind energy report 2017.

ABEEólica. (2023). Infowind #30.

Agrawal, V. V., Toktay, L. B., & Yücel, Ş. (2022). Non-ownership business models for solar energy. *Manufacturing & Service Operations Management, 24*, 2048–2063. https://doi.org/10.1287/msom.2021.1049

Ahlborg, H. (2017). Towards a conceptualization of power in energy transitions. *Environmental Innovation and Societal Transitions, 25*, 122–141. https://doi.org/10.1016/j.eist.2017.01.004

Bellekom, S., Arentsen, M., & Van, G. K. (2016). Prosumption and the distribution and supply of electricity. *Energy, Sustainability and Society, 6*, 1–17. https://doi.org/10.1186/s13705-016-0087-7

Blazquez, J., Fuentes, R., & Manzano, B. (2020). On some economic principles of the energy transition. *Energy Policy, 147*, 111807. https://doi.org/10.1016/j.enpol.2020.111807

Bocken, N. M. P., & Short, S. W. (2016). Towards a sufficiency-driven business model: Experiences and opportunities. *Environmental Innovation and Societal Transitions, 18*, 41–61. https://doi.org/10.1016/j.eist.2015.07.010

Bradshaw, A. (2017). Regulatory change and innovation in Latin America: The case of renewable energy in Brazil. *Utilities Policy, 49*, 156–164. https://doi.org/10.1016/j.jup.2017.01.006

Bukari, D., Kemausuor, F., Quansah, D. A., & Adaramola, M. S. (2021). Towards accelerating the deployment of decentralised renewable energy minigrids in Ghana: Review and analysis of barriers. *Renewable and Sustainable Energy Reviews, 135*, 110408. https://doi.org/10.1016/j.rser.2020.110408

Calderon-Tellez, J. A., Sato, C., Bell, G., & Herrera, M. M. (2023). Project management and system dynamics modelling: Time to connect with innovation and sustainability. *Systems Research and Behavioral Science, 1–27*. https://doi.org/10.1002/sres.2926

Carayannis, E., Kostis, P., Dinçer, H., & Yüksel, S. (2022a). Balanced-scorecard-based evaluation of knowledge-oriented competencies of distributed energy investments. *Energies, 15*, 8245.

Carayannis, E. G., Campbell, D. F. J., & Grigoroudis, E. (2022b). Helix Trilogy: The Triple, quadruple, and quintuple innovation helices from a theory, policy, and practice set of perspectives. *Journal of the Knowledge Economy, 13*, 2272–2301. https://doi.org/10.1007/s13132-021-00813-x

Castaneda, M., Franco, C. J., & Dyner, I. (2017a). Evaluating the effect of technology transformation on the electricity utility industry. *Renewable and Sustainable Energy Reviews, 80*, 341–351. https://doi.org/10.1016/j.rser.2017.05.179

Castaneda, M., Jimenez, M., Zapata, S., et al. (2017b). Myths and facts of the utility death spiral. *Energy Policy, 110*, 105–116. https://doi.org/10.1016/j.enpol.2017.07.063

Castro, M. A. L. (2020). Urban microgrids: Benefits, challenges, and business models. In Noura Guimarães LBT-TR and P of LAET (eds.), *The regulation and policy of Latin American energy transitions* (pp 153–172). Elsevier.

Ce, O. (2009). Benefits of using hybrid business models within a supply chain. *International Journal of Production Economics, 120*, 501–511. https://doi.org/10.1016/j.ijpe.2009.04.006

Chasin, F., Paukstadt, U., & Gollhardt, T. (2020). Smart energy driven business model innovation: An analysis of existing business models and implications for business model change in the energy sector. *Journal of Cleaner Production, 269*, 122083. https://doi.org/10.1016/j.jclepro.2020.122083

Cosenz, F. (2017). Supporting start-up business model design through system dynamics modelling. *Management Decision, 55*, 57–80.

Cosenz, F., & Bivona, E. (2020). Fostering growth patterns of SMEs through business model innovation. A tailored dynamic business modelling approach. *Journal of Business Research.* https://doi.org/10.1016/j.jbusres.2020.03.003

Cosenz, F., Rodrigues, V. P., & Rosati, F. (2019). Dynamic business modeling for sustainability: Exploring a system dynamics perspective to develop sustainable business models. *Business Strategy and the Environment, 1–14*. https://doi.org/10.1002/bse.2395

Dutra, R. M., & Szklo, A. S. (2008). Incentive policies for promoting wind power production in Brazil: Scenarios for the Alternative Energy Sources Incentive Program (PROINFA) under the New Brazilian electric power sector regulation. *Renewable Energy, 33*, 65–76. https://doi.org/10.1016/j.renene.2007.01.013

Eleftheriadis, I. M., & Anagnostopoulou, E. G. (2015). Identifying barriers in the diffusion of renewable energy sources. *Energy Policy, 80*, 153–164. https://doi.org/10.1016/j.enpol.2015.01.039

Farla, J., Markard, J., Raven, R., & Coenen, L. (2012). Sustainability transitions in the making: A closer look at actors, strategies and resources. *Technological Forecasting and Social Change, 79*, 991–998. https://doi.org/10.1016/j.techfore.2012.02.001

Ford, A. (1997). System dynamics and the electric power industry. *System Dynamics Review, 13*, 57–85. https://doi.org/10.1002/(SICI)1099-1727(199721)13:1%3c57::AID-SDR117%3e3.0.CO;2-B

Ford, R., & Hardy, J. (2020). Are we seeing clearly? The need for aligned vision and supporting strategies to deliver net-zero electricity systems. *Energy Policy, 147*, 111902. https://doi.org/10.1016/j.enpol.2020.111902

Forrester, J. W. (1997). Industrial dynamics. *The Journal of the Operational Research Society, 48*, 1037–1041. https://doi.org/10.1057/palgrave.jors.2600946

Gronum, S., Steen, J., & Verreynne, M.-L. (2015). Business model design and innovation: Unlocking the performance benefits of innovation. *Australian Journal of Management, 41*, 585–605. https://doi.org/10.1177/0312896215587315

Gsodam, P., Rauter, R., & Baumgartner, R. J. (2015). The renewable energy debate: How Austrian electric utilities are changing their business models. *Energy, Sustainability and Society, 5*,. https://doi.org/10.1186/s13705-015-0056-6

Hamelink, M., & Opdenakker, R. (2019). How business model innovation affects firm performance in the energy storage market. *Renewable Energy, 131,* 120–127. https://doi.org/10.1016/j.renene.2018.07.051

Hekkert, M., & Negro, S. (2009). Functions of innovation systems as a framework to understand sustainable technological change: Empirical evidence for earlier claims. *Technological Forecasting and Social Change, 76,* 584–594. https://doi.org/10.1016/j.techfore.2008.04.013

Hekkert, M., Suurs, R., Negro, S., et al. (2007). Functions of innovation systems: A new approach for analysing technological change. *Technological Forecasting and Social Change, 74,* 413–432. https://doi.org/10.1016/j.techfore.2006.03.002

Herrera, M. M., & Trujillo-Díaz, J. (2022). Towards a strategic innovation framework to support supply chain performance. *International Journal of Productivity and Performance Management, 71,* 1872–1894. https://doi.org/10.1108/IJPPM-03-2020-0131

Herrera, M. M., Dyner, I., & Cosenz, F. (2018). *Alternative energy policy for mitigating the asynchrony of the wind-power industry's supply chain in Brazil.*

Herrera, M. M., Dyner, I., & Cosenz, F. (2019). Assessing the effect of transmission constraints on wind power expansion in northeast Brazil. *Utilities Policy, 59,.* https://doi.org/10.1016/j.jup.2019.05.010

Herrera, M. M., Cosenz, F., & Dyner, I. (2020a). Blending collaborative governance and dynamic performance management to foster policy coordination in renewable energy supply chains. In C. Bianchi, L. Luna-Reyes, & E. Rich (eds.), *Enabling collaborative governance through systems modeling methods. System dynamics for performance management & governance* (pp 237–261). Springer, Cham.

Herrera, M. M., Dyner, I., & Cosenz, F. (2020b). Benefits from energy policy synchronisation of Brazil's North-Northeast interconnection. *Renewable Energy, 162,* 427–437. https://doi.org/10.1016/j.renene.2020.08.056

Herrera, M. M., Dyner, I., & Uriona Maldonado, M. (2020c). Modelling the wind supply chain to reduce emissions: How could affect transmission congestion? *Dynamics of Energy, Environment and Economy: A Sustainability Perspective,* 187–205.

Hirt, L. F., Schell, G., Sahakian, M., & Trutnevyte, E. (2020). A review of linking models and socio-technical transitions theories for energy and climate solutions. *Environmental Innovation and Societal Transitions, 35,* 162–179. https://doi.org/10.1016/j.eist.2020.03.002

Holtz, G., Alkemade, F., De Haan, F., et al. (2015). Prospects of modelling societal transitions: Position paper of an emerging community. *Environmental Innovation and Societal Transitions, 17,* 41–58. https://doi.org/10.1016/j.eist.2015.05.006

Horváth, D., & Szabó, R. Z. (2018). Evolution of photovoltaic business models: Overcoming the main barriers of distributed energy deployment. *Renewable and Sustainable Energy Reviews, 90*, 623–635. https://doi.org/10.1016/j. rser.2018.03.101

Huijben, J. C. C. M., & Verbong, G. P. J. (2013). Breakthrough without subsidies? PV business model experiments in the Netherlands. *Energy Policy, 56*, 362–370. https://doi.org/10.1016/j.enpol.2012.12.073

Jacobsson, S., & Karltorp, K. (2013). Mechanisms blocking the dynamics of the European offshore wind energy innovation system – Challenges for policy intervention. *Energy Policy, 63*, 1182–1195. https://doi.org/10. 1016/j.enpol.2013.08.077

Jimenez, M., Franco, C. J., & Dyner, I. (2016). Diffusion of renewable energy technologies: The need for policy in Colombia. *Energy, 111*, 818–829. https:/ /doi.org/10.1016/j.energy.2016.06.051

Karami, M., & Madlener, R. (2021). Business model innovation for the energy market: Joint value creation for electricity retailers and their customers. *Energy Research & Social Science, 73*, 101878. https://doi.org/10.1016/j.erss.2020. 101878

Li, C., & Shen, B. (2019). Accelerating renewable energy electrification and rural economic development with an innovative business model: A case study in China. *Energy Policy, 127*, 280–286. https://doi.org/10.1016/j.enpol.2018. 12.009

Lozano, R., & Reid, A. (2018). Socially responsible or reprehensible? Investors, electricity utility companies, and transformative change in Europe. *Energy Research & Social Science, 37*, 37–43. https://doi.org/10.1016/j.erss.2017. 09.018

Markard, J., Raven, R., & Truffer, B. (2012). Sustainability transitions: An emerging field of research and its prospects. *Research Policy, 41*, 955–967. https://doi.org/10.1016/j.respol.2012.02.013

Mastropietro, P., Batlle, C., Barroso, L. A., & Rodilla, P. (2014). Electricity auctions in South America: Towards convergence of system adequacy and RES-E support. *Renewable and Sustainable Energy Reviews, 40*, 375–385. https://doi.org/10.1016/j.rser.2014.07.074

McDowall, W., & Geels, F. W. (2017). Ten challenges for computer models in transitions research: Commentary on Holtz et al. *Environmental Innovation and Societal Transitions, 22*, 41–49. https://doi.org/10.1016/j.eist. 2016.07.001

Morcillo, J. D., Franco, C. J., & Angulo, F. (2017). Delays in electricity market models. *Energy Strategy Reviews, 16*, 24–32. https://doi.org/10.1016/j.esr. 2017.02.004

Pereira, A. O., Pereira, A. S., La Rovere, E. L., et al. (2011). Strategies to promote renewable energy in Brazil. *Renewable and Sustainable Energy Reviews, 15*, 681–688. https://doi.org/10.1016/j.rser.2010.09.027

Petzer B. J. M. B., Wieczorek, A. A., & Verbong, G. G. (2020). Cycling as a service assessed from a combined business-model and transitions perspective. *Environmental Innovation and Societal Transitions, 36*, 255–269. https://doi.org/10.1016/j.eist.2019.09.001

Porrua, F., Bezerra, B., Barroso, L. A., et al. (2010). *Wind power insertion through energy auction in Brazil*. In Power and Energy Society General Meeting. IEEE, pp. 1–8.

Rehman, F. U., Islam, M. M., Ullah, M., et al. (2023). Information digitalization and renewable electricity generation: Evidence from South Asian countries. *Energy Reports, 9*, 4721–4733. https://doi.org/10.1016/j.egyr.2023.03.112

Richter, M. (2013a). Business model innovation for sustainable energy: German utilities and renewable energy. *Energy Policy, 62*, 1226–1237. https://doi.org/10.1016/j.enpol.2013.05.038

Richter, M. (2013b). German utilities and distributed PV: How to overcome barriers to business model innovation. *Renewable Energy, 55*, 456–466. https://doi.org/10.1016/j.renene.2012.12.052

Ruggiero, S., Varho, V., & Rikkonen, P. (2015). Transition to distributed energy generation in Finland: Prospects and barriers. *Energy Policy, 86*, 433–443. https://doi.org/10.1016/j.enpol.2015.07.024

Ruggiero, S., Kangas, H. L., Annala, S., & Lazarevic, D. (2021). Business model innovation in demand response firms: Beyond the niche-regime dichotomy. *Environmental Innovation and Societal Transitions, 39*, 1–17. https://doi.org/10.1016/j.eist.2021.02.002

Sadjadi, E. N., & Fernández, R. (2023). Relational marketing promotes sustainable consumption behavior in renewable energy production. *Sustainability, 15*, 5714. https://doi.org/10.3390/su15075714

Sarasini, S., & Linder, M. (2018). Integrating a business model perspective into transition theory: The example of new mobility services. *Environmental Innovation and Societal Transitions, 27*, 16–31. https://doi.org/10.1016/j.eist.2017.09.004

Sterman, J. D. (2000). *Business dynamics systems thinking and modeling for a complex world*. McGraw-Hill.

Sterman, J. D. (2003). System dynamics: Systems thinking and modeling for a complex world. *European Journal of Computer Science and Information, 21*, 35–39.

Su, C.-W., Pang, L.-D., Qin, M., et al. (2023). The spillover effects among fossil fuel, renewables and carbon markets: Evidence under the dual dilemma of climate change and energy crises. *Energy, 274*, 127304. https://doi.org/10.1016/j.energy.2023.127304

Suurs, R. A. A. (2009). *Motors of sustainable innovation: Towards a theory on the dynamics of technological innovation systems.*

Trotter, P. A., & Brophy, A. (2022). Policy mixes for business model innovation: The case of off-grid energy for sustainable development in sub-Saharan Africa. *Research Policy, 51*, 104528. https://doi.org/10.1016/j.respol.2022.104528

Uriona, M., & Grobbelaar, S. S. (2019). Innovation system policy analysis through system dynamics modelling: A systematic review. *Science and Public Policy, 46*, 28–44. https://doi.org/10.1093/scipol/scy034

Van Den Bergha, J. C. J. M. (2012). Editorial; EIST one year: Something to celebrate? *Environmental Innovation and Societal Transitions, 4*, 1–6. https://doi.org/10.1016/j.eist.2012.07.002

Visintainer, L., Gerstlberger, W., Ferreira, M., & Frank, A. G. (2021). Energy Research & Social Science How governments, universities, and companies contribute to renewable energy development? A municipal innovation policy perspective of the triple helix. *Energy Research & Social Science, 71*, 101854. https://doi.org/10.1016/j.erss.2020.101854

Walrave, B., & Raven, R. (2016). Modelling the dynamics of technological innovation systems. *Research Policy, 45*, 1833–1844. https://doi.org/10.1016/j.respol.2016.05.011

Weigelt, C., Lu, S., & Verhaal, J. C. (2021). Blinded by the sun: The role of prosumers as niche actors in incumbent firms' adoption of solar power during sustainability transitions. *Research Policy, 50*, 104253. https://doi.org/10.1016/j.respol.2021.104253

Xi, Y., Huynh, A. N. Q., Jiang, Y., & Hong, Y. (2023). Energy transition concern: Time-varying effect of climate policy uncertainty on renewables consumption. *Technological Forecasting and Social Change, 192*, 122551. https://doi.org/10.1016/j.techfore.2023.122551

Zapata, S., Castaneda, M., Herrera, M. M., & Dyner, I. (2023). Investigating the concurrence of transmission grid expansion and the dissemination of renewables. *Energy, 127571,*. https://doi.org/10.1016/j.energy.2023.127571

A Macro Perspective of the Innovation Process: A Dynamic Performance Management Approach

Ricardo E. Buitrago R. ⓘD

Abstract This chapter introduces a novel approach to macroeconomic policy design and analysis, using Dynamic Performance Management (DPM) to investigate the complex relationship between institutional quality and innovation. Integrating performance management, system dynamics, and political economy enhances performance reports, governance, and policy design by mapping output outcome measures and performance drivers. A DPM model is proposed, highlighting the interaction between strategic resources, performance drivers, and final outcomes, focusing on taxation's impact on institutional quality. This approach aims to improve the understanding of resources, drivers, and outcomes needed for effective public policy in national innovation systems. Although the preliminary model centres on one variable, it can be expanded to include

R. E. Buitrago R. (✉)
EGADE Business School, Tecnológico de Monterrey, México City, Mexico
e-mail: ricardo.buitrago@tec.mx

© The Author(s), under exclusive license to Springer Nature
Switzerland AG 2023
M. M. Herrera (ed.), *Business Model Innovation for Energy Transition*,
Palgrave Studies in Democracy, Innovation, and Entrepreneurship
for Growth, https://doi.org/10.1007/978-3-031-34793-1_2

others, providing valuable insights for policymaking, and fostering institutional collaboration. Further research on combining qualitative modelling with simulation for a robust macroeconomic policymaking methodology is recommended.

Keywords Dynamic Performance Management · Institutional · System Dynamics · Economic · Innovation

2.1   INTRODUCTION

Researchers, specialists, and policymakers acknowledge the significance of innovation for economic growth and societal well-being as a whole. Previous research has employed conventional econometric models and variables to assess the effect of institutions on innovation (Buitrago R. et al., 2021). To close gaps and expand on past research, this chapter examines the impact of various institutional conditions (exogenous) on the innovation (patents) as generator of economic growth. This chapter examines the nonlinear interactions in the ability of economies to produce innovation. We follow the Dynamic Performance Management (DPM) approach (Bianchi, 2016; Cosenz & Noto, 2014) to conduct this analysis.

Due to the similarities between institutional economics and systems dynamics, we choose to perform this study using both approaches. According to Kapp (1976), the most fundamental principle of institutional economics is Gunnar Myrdal's (1957) concept of cyclical and cumulative causation, which is believed to explain economic systems' nonequilibrium dynamics.

System dynamics employs a method of analysis that is strikingly similar to the approach employed by institutional economics in the sense that both approaches use pattern modelling (Radzicki, 1988, 2004; Radzicki & Tauheed, 2009). The system dynamics approach does not seek to represent systems; rather, it aims to model problems from a systems perspective. As with institutional economics, system dynamics uses a wide variety of accessible data to develop a pattern or explanation (i.e., a simulation model) for a given scenario. System dynamics modelling is an iterative process in which the process steps (including identifying relevant information) are routinely revisited and the model altered, as the modelling process itself creates new insights into the problem (Bianchi, 2016, 2022; Fratesi, 2010; Radzicki, 2021).

Based on the literature review, we can state that this work is the first approach to study the nonlinear relationships in innovation using DPM, extending its scope to the field of international political economy by the use and combination of alternative approaches to explain the proposed causal relationships (Bianchi, 2016, 2022; Buitrago R. et al., 2021; Cosenz & Noto, 2014, 2016). An important outcome of innovation research is the identification of the feedback loop between innovation rates and the national economy, in which innovation's impact on economic growth is in turn influenced by national prosperity. Nonetheless, this relationship between innovation and external (institutional) factors is understudied and cannot be fully explained by conventional cognitive analysis. This study addresses this research gap.

This chapter is organized as follows: Sect. 2.2 summarizes the literature review and construction of causal links; Sect. 2.3 covers the methodological structure in detail; and Sect. 2.4 summarizes the findings and comments, limits, and suggests future study directions.

2.2   Literature Review

Economic diversification (complexity) is defined as the evolution of an economy's composition and quality of the economic sectors. It is both a cause and effect of increased production and revenue (Hidalgo & Hausmann, 2009; Hidalgo et al., 2007; Saviotti, 1996). Economic diversification alters the available choices in an economy, from jobs and vocational options to consumption patterns. Together with institutional and technological improvements, it enables the economy to diversify into other areas. As stated by Kuznets (1971, p. 1):

A country's economic growth may be defined as a long-term rise in capacity to supply increasingly diverse economic goods to its population, this growing capacity based on advancing technology and the institutional and ideological adjustments that it demands. All three components of the definition are important.

The central tenet of this "complexity" approach is that each country is defined by unique fundamental endowments, dubbed capabilities, which encompass all economic resources and the attributes of a country's societal structure that enable the same country to produce and export a basket of tradeable commodities. These capabilities are non-tradeable and, in some cases, difficult to quantify and compare (Cristelli et al., 2013).

The conventional approach to economic performance analysis is based on a country's endowments of physical and human capital, labour, and natural resources, as well as the general quality of its institutions, which serve as the foundation for determining relative costs and associated patterns of specialization (Hausmann et al., 2007).

Economic growth is greatly reliant on innovation (Gordon, 2004). Technological innovation and development enable economic progress. Through invention and creativity, as well as foreign technology absorption, new or improved technology can be developed. Allowing for such technological advancements necessitates the establishment of supportive institutions and regulations. This suggests that an economy's competitiveness is contingent upon the effectiveness of government policy (Lim & Moon, 2004). Economic growth is determined by the degree to which institutions and systemic variables encourage technical advancements (OECD, 2001, 2003). If firms are to be incentivized to innovate, they must have the capacity to appropriate at least a portion of the value created by their innovations. New knowledge production can be viewed as an evolutionary process; evolutionary theory describes how new concepts emerge through the variation and mutation of existing and established solutions (Milling, 2002). In contrast, the benefits of an innovation to the economy as a whole are highly dependent on the extent to which the new knowledge associated with it is made available for others to use and build upon (Cohen et al., 2002).

Technology and human capital are inextricably linked, indispensable, and interdependent. A significant portion of technological advancement is the outcome of investment in human capital. Without competent employees, machines, equipment, scientific instruments, and the legal and financial systems, most of the contemporary civilization would not function. To advance technology, it is vital to recruit and retain qualified workers. In addition, society needs technical and managerial abilities to make the most use of technology and human capital (Warhuus & Basaiawmoit, 2014; Winkler et al., 2015).

In this line, it's essential to address the causes of "brain drain" as an obstacle to producing innovation. The asymmetry between a nation's capacity to produce numbers of highly trained personnel and its capacity to absorb them is a significant component of internal push forces, more so than poverty or underdevelopment. The emigration of skilled professionals (researchers) results from international imbalances that allow advanced industrial nations to offer more attractive remunerations, work facilities, social standing, and general living conditions to those whose skills and talents they require. The internal structural imbalances between the supply of researchers produced by a society's educational system and the demand for their services within that society is another source of brain drain. Finally, brain drain is the result of individual differences relating to, among other things, past training and accomplishments, current situation, and the individual's surrounding social network (Brock & Blake, 2015; Lister, 2017; Perrou & Savvaidou, 2019; Pescaru, 2014; Portes, 1976).

Institutions of the modern economy must be taken into account while considering economic progress and welfare. North (1986) maintains that robust, dependable institutions are necessary for the current economic system to function properly. While certain institutions are more established than others, most institutions in developing countries are still in their development. The country's lack of institutional development has been cited as a source of macroeconomic volatility, which may be explained by the adverse effects on economic growth and prosperity (Acemoglu, 2003; Acemoglu et al., 2001; Acemoglu & Johnson, 2003; Acemoglu & Robinson, 2012; Hnatkovska & Loayza, 2005; Ramey & Ramey, 1995).

Thus, we sought to understand how various institutional characteristics promote and prevent economic complexity. Due to the firm's contact with a diverse spectrum of stakeholders, the institutional framework in which it operates is critical. In a business setting, regulatory and normative factors influence how firms behave (North, 1990; Peng & Heath, 1996). Economic outcomes and internationalization are determined by factors such as government stability, political parties, the predictability of the legal system, quantity, allocation of available resources, and contractual enforcement (Besley et al., 2010; Blume et al., 2009; Buitrago R. et al., 2021; Cuervo-Cazurra, 2016; Cuervo-Cazurra & Alfonso Dau, 2009; Rodrik, 1999).

The legal infrastructure of a country's capacity to resolve disputes and enforce contracts encourages businesses to rely on it (Li, 2009). According to Kramer (1999), rules are predicated on the capability of institutions to forecast their own behaviour. At the country level, trust in a country's laws is reflected in confidence in the country's legal system (Lin & Wang, 2008; Muethel & Bond, 2013).

The productive structure of a country is determined not only by its factor endowment but also by its social capital and the quality of its institutions. According to previous research, the complexity and diversity of items exported by a country are a solid predictor of the economy's resources. Complex products (innovative) demand a more significant amount of tacit knowledge and involve a greater amount of distributed information than products based on resource abundance or low labour costs (Hausmann et al., 2007; Hidalgo & Hausmann, 2009; Sheng & Yang, 2016; Zhu & Fu, 2013).

Prior research on the significance of productive structures generated a range of measures of technical sophistication (Dosi, 1991). Other quantitative attempts rely on iterative or dimensionality reduction approaches (Fleming & Sorenson, 2001), and others on variables that were averaged over other indicators, such as patent, human capital, or income data (Archibugi & Coco, 2004; Desai et al., 2010; Hausmann et al., 2007; Lall, 2003).

2.3   METHODOLOGY

This research aims to create a DPM model to illustrate the dynamic behaviour of economic performance bounded by factor endowments, Foreign Direct Investment (FDI), and institutional conditions, to assist researchers and policymakers in gaining a deeper understanding of the economic complexity system.

2.3.1   Principles of System Dynamics Modelling

System dynamics modelling is based on a number of fundamental systems principles at the most general level. Among the most critical principles are the following:

- Accumulation principle—this theory states that all dynamic behaviour in the world arises as a result of flows accumulating in stocks.
- The notion of ubiquitous feedback says that stocks and flows do not exist in isolation but are virtually always a component of feedback loops.

- Structure determines behaviour, which says that to alter the behaviour of a system, the system's structure must be changed.

2.3.1.1 *Dynamic Performance Management*

Dynamic performance management (DPM) combines performance management and system dynamics (Bianchi & Rivenbark, 2012; Bianchi & Tomaselli, 2015; Cosenz & Noto, 2014; Noto, 2017). This approach enables the identification and comprehension of desired end-results to develop a small but meaningful set of performance indicators (drivers). These drivers serve as strategic levers to close the gap between existent and envisaged results. Managers must accumulate, protect, and use a sufficient endowment of strategic resources to influence such drivers. The feedback loops that underpin the dynamics of various strategic resources entail that the flows affecting them are time-dependent and monitored across a time lag. The results are modelled as flows (in or out) that alter the stocks of strategic resources over a certain time due to decision-makers' activities.

As indicated previously, we chose the system dynamics technique since it can establish the system's operational status through a causal loop diagram design, to identify the causal relationship between the model's variables. If the result of the effect is positive (+), it is called a reinforcing loop; if the result is negative (−), it is a balancing loop. Also, constructing a DPM chart helps identify the system's critical resources, performance drivers, and end-results.

According to Radzicki (2021), complex socioeconomic systems are challenging to comprehend and regulate due to a variety of structural characteristics. These features include the following:

- Components embedded in complex networks of interconnected feedback loops, where time and space divide and blur cause-and-effect interactions, making it difficult to explain the relationship between system structure and behaviour.
- Structures that typically emerge through evolutionary forces rather than deliberate design, which means they frequently lack robust architectures that can mitigate their vulnerability to major external shocks that can significantly disrupt their normal behaviour.

• Feedback structures that frequently result in robust and persistent undesirable behaviour.

2.4   RESULTS

To depict this complexity, this study uses the following structure to explain innovation process in emerging economies:

Conceptualization using the DPM chart: Bianchi (2010, 2012) indicates that social systems can be articulated in terms of strategic resources (resources owned by the entire system), end-results (what is desired or required to accomplish), and performance indicators (intermediate results that explain how to employ the strategic resources in order to achieve the end-results). Strategic resources can be thought of as variables that are subject to accumulation/depletion processes (stock variables). End-results are frequently expressed as "'flows"; the process by which stock variables change over time, as illustrated in Fig. 2.1.

Figure 2.2 depicts the end-results through the sequential levels of (1) change in institutional quality, (2) change in FDI, (3) change in

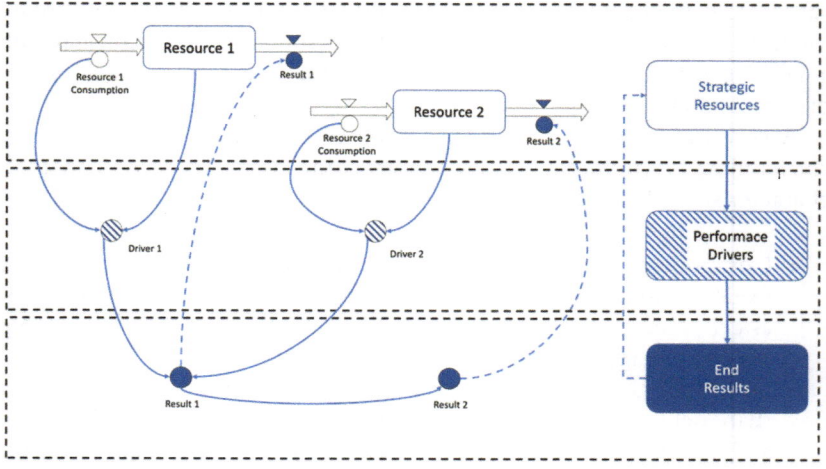

Fig. 2.1 Dynamic performance management perspective—DPM (*Source* Dynamic performance management perspective [Bianchi, 2012])

university-industry collaboration, (4) change in available research and training services, (5) human capital formation, (6) brain drain, (7) change in registered patents (CRP), (8) change in approved patents (CAP), (9) change in used patents (CUP), and (10) change in Real GDP.

These final outcomes are influenced by performance drivers. Figure 2.2 depicts the following performance drivers: Cost Of Enforcement Contracts Ratio, Cost Of Cash Repatriation Ratio, Diversion Of Public Funds Ratio, Taxation Ratio, Institutional Quality Ratio, R&D Joint Investment Ratio, R&D Business Investment Ratio, R&D Government Investment Ratio, Scientific Infrastructure Ratio, University-Industry Collaboration Projects Ratio, Human Capital Ratio, Registered Patents Ratio, Approved These ratios are computed by comparing the current state to the desired state through policy design or benchmark. Figure 2.2 also depicts how "qualitative system dynamics based on stocks and flows" (Wolstenholme, 1999, p. 423) can contribute to performance management and governance. This perspective on modelling borrows from qualitative modelling to enrich this field of research and practice. This paper does not claim to depict complex cause-and-effect relationships that can be transformed into a simulation model for policy design in the absence of additional data. In the context of this work, the primary function of DPM is to map output outcome measures and the performance drivers

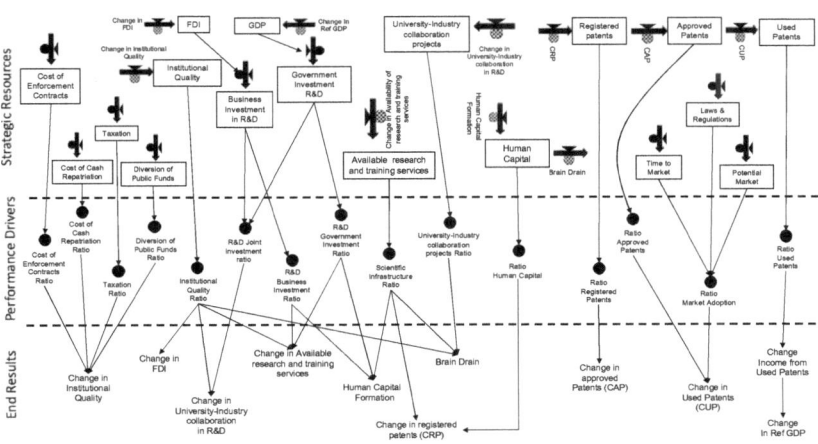

Fig. 2.2 DPM innovation (*Source* Author's elaboration)

affecting them. Therefore, the analysis conducted in this paper aims to connect three traditionally separate fields of research and practice (namely, performance management and governance, system dynamics, and political economy). These disciplines typically employ distinct methods and instruments for analysis. The use of systems approaches in outcome-based performance management, leading to the identification of causal relationships between variables affecting results over time, can enhance the quality of performance reports, governance, accountability, and policy design. This is a preliminary stage for implementing simulation modelling as a potential next stage of analysis to enhance decision-making. The gradual introduction of mapping approaches in outcome-based performance management, to illustrate causal relationships between variables affecting results over time, may improve the quality of performance reports and, consequently, policy design. In turn, this would increase awareness in the field of simulation's potential to further enhance dynamic performance management.

Finally, due to the complexity of the DPM, it is necessary to split it in subsystems that allow to propose an initial model in the form of a causal loop diagram (CLD), as shown in Fig. 2.3. This CLD shows two balancing loops and two reinforcing loops, depicting the interaction between the strategic resources, performance drivers, and final outcomes. This first approach is focused on the effects of *Taxation* on *Institutional Quality* and how this affects the other indicators in the system.

The first balancing loop is related with the need of researchers to increase the human capital, there's a gap between the number of available researchers and the desired number of researchers. The first reinforcing loop is showing how institutional quality which increases university-industry collaboration, which increases human capital, which increases the patents, which increases reference GDP, which increases the R&D budget, which decreases the R&D budget gap, which decreases taxation, which decreases institutional quality. The second balancing loop is R&D budget, which decreases the R&D budget gap, which decreases taxation, which increases R&D Budget. Finally, the second reinforcing loop is R&D budget which increases desired researchers, which increases human capital, which increases the patents, which increases reference GDP, which increases the R&D budget.

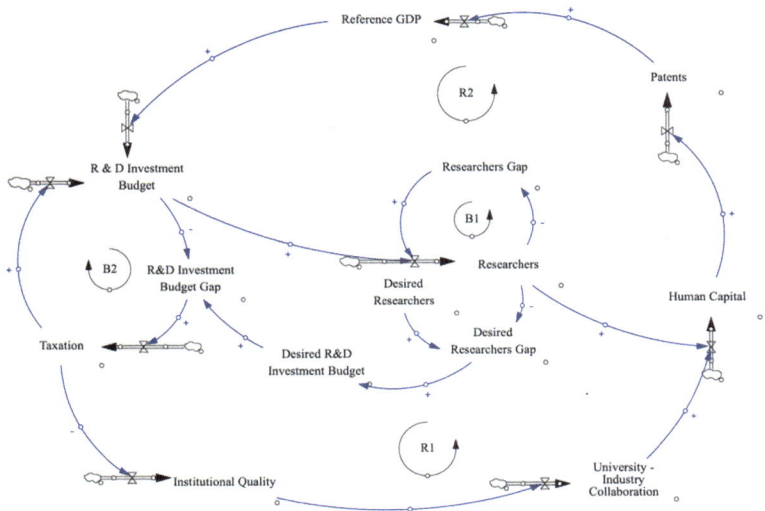

Fig. 2.3   Causal loop diagram

2.5   Conclusions

As mentioned before this is a first approach to model a macroeconomic environment regarding institutional quality and innovation, measured by the patents that can be created and used in the system. This approach aims to show the complexity of the innovation process as it's seen from a macroeconomic point of view. National innovation systems are driven by public policy, this DPM proposal could help in the understanding of the resources, drivers, and end-results required to design policies that positively impact the innovation on a determined economy.

This work has examined how DPM can be utilized to analyse macroeconomic issues, such as institutional quality in this instance. Such an approach contributes to bridging the gap between conventional economic analysis methods and the dynamic complexity that characterizes policymaking. A number of insights have emerged for reframing the policymaking process. A DPM approach can be useful for altering how policymakers view macroeconomic issues and overcoming collaboration barriers. In fact, it can enable each institution in a region to identify their strategic resources, performance drivers, and innovation-related outcomes (in this case).

In addition, this methodology assists public agencies in understanding that long-term performance can be evaluated in relation to the outcomes that public policy will produce. Although this proposal focuses on a single variable of institutional quality, it can be expanded to include other variables deemed pertinent to the analysis. Despite the fact that I believe the case analysis has demonstrated the usefulness of the proposed method for enhancing policymaking, I am aware that additional field research will be required to combine qualitative modelling with simulation in order to improve a DPM approach to macroeconomic policymaking.

REFERENCES

Acemoglu, D. (2003). Root causes. *Finance and Development, 40*(2), 26–30.

Acemoglu, D., & Johnson, S. (2003). Institutions, corporate governance. *Corporate Governance and Capital Flows in a Global Economy, 1,* 327.

Acemoglu, D., Johnson, S., & Robinson, J. A. (2001). The colonial origins of comparative development: An empirical investigation. *The American Economic Review, 91*(5), S0022050701228113. https://doi.org/10.1017/S00220507 01228113

Acemoglu, D., & Robinson, J. A. (2012). *Why nations fail: The origins of power, prosperity and poverty*. Crown Publishers.

Archibugi, D., & Coco, A. (2004). A new indicator of technological capabilities for developed and developing countries (ArCo). *World Development, 32*(4), 629–654. Pergamon. https://doi.org/10.1016/J.WOR LDDEV.2003.10.008

Besley, T., Persson, T., & Sturm, D. M. (2010). Political competition, policy and growth: Theory and evidence from the US. *Review of Economic Studies.* https://doi.org/10.1111/j.1467-937X.2010.00606.x

Bianchi, C. (2010). Improving performance and fostering accountability in the public sector through system dynamics modelling: From an 'external' to an 'internal' perspective. *Systems research and behavioral science, 27*(4), 361–384. John Wiley & Sons, Ltd. https://doi.org/10.1002/SRES.1038

Bianchi, C. (2012). Enhancing performance management and sustainable organizational growth through system-dynamics modelling. In *Systemic management for intelligent organizations* (pp. 143–161). Springer. https://doi.org/10. 1007/978-3-642-29244-6_8

Bianchi, C. (2016). Dynamic performance management. *Springer.* https://doi. org/10.1007/978-3-319-31845-5

Bianchi, C. (2022). Enhancing policy design and sustainable community outcomes through collaborative platforms based on a dynamic performance management and governance approach. In B. G. Peters & G. Fontaine (Eds.), *Handbook of policy design* (pp. 407–429). Edward Elgar.

Bianchi, C., & Rivenbark, W. (2012). A comparative analysis of performance management systems: The cases of Sicily and North Carolina. *Public Performance and Management Review, 35*(3), 509–526. https://doi.org/10.2753/PMR1530-9576350307

Bianchi, C., & Tomaselli, S. (2015). A dynamic performance management approach to support local strategic planning. *International Review of Public Administration, 20*(4), 370–385. https://doi.org/10.1080/12294659.2015.1088687

Blume, L., Müller, J., Voigt, S., et al. (2009). The economic effects of constitutions: Replicating-and extending-persson and tabellini. *Public Choice.* https://doi.org/10.1007/s11127-008-9389-4

Brock, G., & Blake, M. (2015). *Debating brain drain: May governments restrict emigration?* Oxford University Press.

Buitrago, R. R. E., Barbosa Camargo, M. I., & Cala Vitery, F. (2021). Emerging economies' institutional quality and international competitiveness: A PLS-SEM approach. *Mathematics, 9*(9), 928. https://doi.org/10.3390/math9090928

Cohen, W. M., Goto, A., Nagata, A., et al. (2002). RandD spillovers, patents and the incentives to innovate in Japan and the United States. *Research Policy, 31*(8–9), 1349–1367. https://doi.org/10.1016/S0048-7333(02)00068-9

Cosenz, F., & Noto, G. (2014). A dynamic simulation approach to frame drivers and implications of corruption practices on firm performance. *European Management Review, 11*(3–4), 239–257. https://doi.org/10.1111/emre.12037

Cosenz, F., & Noto, G. (2016). Applying system dynamics modelling to strategic management: A literature review. *Systems Research and Behavioral Science, 33*(6), 703–741. https://doi.org/10.1002/sres.2386

Cristelli, M., Gabrielli, A., & Tacchella, A. et al. (2013). Measuring the intangibles: A metrics for the economic complexity of countries and products. *PLOS ONE 8*(8), e70726. Public Library of Science. https://doi.org/10.1371/JOURNAL.PONE.0070726

Cuervo-Cazurra, A. (2016). Corruption in international business. *Journal of World Business, 51*(1), 35–49. Elsevier Inc. https://doi.org/10.1016/j.jwb.2015.08.015

Cuervo-Cazurra, A., & Alfonso Dau, L. (2009). *Multinationalization in response to reforms.* In 2009. https://www.scopus.com/inward/record.uri?eid=2-s2.0-84858387015&partnerID=40&md5=e37f5236fb4b14ce9164f472a18d7975

Desai, M., Fukuda-Parr, S., & Johansson, C. et al. (2010). Measuring the technology achievement of nations and the capacity to participate in the network age. *Journal of Human Development, 3*(1), 95–122. Taylor & Francis Group. https://doi.org/10.1080/14649880120105399

Dosi, G. (1991). Diffusion of technologies and social behavior. In N. Nakicenovic & A. Grübler (Eds.), *Diffusion of technologies and social behavior* (pp. 179–208). Springer. https://doi.org/10.1007/978-3-662-02700-4

Fleming, L., & Sorenson, O. (2001). Technology as a complex adaptive system: Evidence from patent data. *Research Policy 30*(7), 1019–1039. https://doi.org/10.1016/S0048-7333(00)00135-9

Fratesi, U. (2010). Regional innovation and competitiveness in a dynamic representation. *Journal of Evolutionary Economics, 20*(4), 515–552. https://doi.org/10.1007/s00191-009-0169-1

Gordon, R. J. (2004). *Five puzzles in the behavior of productivity, investments and innovation* (Nber Working Paper Series).

Hausmann, R., Hwang, J., & Rodrik, D. (2007). What you export matters. *Journal of Economic Growth, 12*, 1–25. https://doi.org/10.2139/ssrn.896243

Hidalgo, C. A., & Hausmann, R. (2009). The building blocks of economic complexity. In Proceedings of the National ..., 1 January 2009. http://www.pnas.org/content/106/26/10570.short

Hidalgo, C. A., Winger, B., & Barabási, A. L., et al. (2007). The product space conditions the development of nations. *Science, 317*(5837), 482–487. American Association for the Advancement of Science. https://doi.org/10.1126/SCIENCE.1144581/SUPPL_FILE/HIDALGO.SOM.PDF.

Hnatkovska, V., & Loayza, N. (2005). Volatility and growth. In J Aizenman & B Pinto (Eds.), *Managing economic volatility and crises: A practitioner's guide* (pp. 65–100). Cambridge University Press. https://doi.org/10.1017/CBO9780511510755.005

Kapp, K. W. (1976). The nature and significance of institutional economics. *Kyklos, 29*(2), 209–232. John Wiley & Sons, Ltd. https://doi.org/10.1111/J.1467-6435.1976.TB01971.X

Kramer, R. M. (1999). Trust and distrust in organizations: Emerging perspectives, enduring questions. *Annual Review of Psychology.* https://doi.org/10.1146/annurev.psych.50.1.569

Kuznets, S. (1971). Modern economic growth: Findings and reflections. https://www.nobelprize.org/prizes/economic-sciences/1971/kuznets/lecture/. Accessed 3 May 2022.

Lall, S. (2003). Indicators of the relative importance of IPRs in developing countries. *Research Policy, 32*(9), 1657–1680. North-Holland. https://doi.org/10.1016/S0048-7333(03)00046-5

Li, S. (2009). Managing international business in relation-based versus rule-based countries. https://doi.org/10.4128/9781606490853

Lim, W. H., & Moon, J. T. (2004). Korea as a knowledge economy: Assessment and lessons. In *Designing new economic framework*. Korea Development Institute (KDI).

Lin, X., & Wang, C. L. (2008). Enforcement and performance: The role of ownership, legalism and trust in international joint ventures. *Journal of World Business*. https://doi.org/10.1016/j.jwb.2007.11.005

Lister, M. J. (2017). A tax-credit approach to addressing brain drain. *Saint Louis University Law Journal Volume, 62*(1), 73–84. https://papers.ssrn.com/sol3/papers.cfm?abstract_id=3226855

Milling, P. M. (2002). Understanding and managing innovation processes. *System Dynamics Review, 18*(1), 73–86. https://doi.org/10.1002/sdr.231

Muethel, M., & Bond, M. H. (2013). National context and individual employees' trust of the out-group: The role of societal trust. *Journal of International Business Studies*. https://doi.org/10.1057/jibs.2013.9

Myrdal, G. (1957). The principle of circular and cumulative causation. In G. Myrdal (Ed.), *Economic theory and underdeveloped regions* (pp. 11–22). Methuen and Co., Ltd.

North, D. C. (1986). The new institutional economics. *Journal of Institutional and Theoretical Economics (JITE), 142*(1), 230–237.

North, D. C. (1990). A transaction cost theory of politics. *Journal of Theoretical Politics, 2*(4), 355–367.

Noto, G. (2017). Combining system dynamics and performance management to support sustainable Urban transportation planning. *Journal of Urban and Regional Analysis, 9*(1), 51–71.

OECD. (2001). *The new economy: Beyond the hype*.

OECD. (2003). *The source of economic growth in OECD countries*.

Peng, M. W., & Heath, P. S. (1996). The growth of the firm in planned economies in transition: Institutions, organizations, and strategic choice. *Academy of Management Review*. https://doi.org/10.5465/AMR.1996.9605060220

Perrou, K., & Savvaidou, K. (2019). Brain drain: The impact on taxation and measures to combat the brain drain. *Annals of the Faculty of Law in Belgrade International Edition*, 238–248. https://doi.org/10.5937/AnaliPFBI904238P

Pescaru, C.-M. (2014). The brain drain phenomenon in Romania: Characteristics, implication. *Revista Universitara De Sociologie, 2*, 40–45.

Portes, A. (1976). Determinants of the brain drain. *International Migration Review, 10*(4), 489–508. https://doi.org/10.1177/019791837601000402

Radzicki, M. J. (1988). Institutional dynamics: An extension of the institutionalist approach to socioeconomic analysis. *Journal of Economic Issues, 22*(3), 633–665. https://doi.org/10.1080/00213624.1988.11504801

Radzicki, M. J. (2004). *Institutional economics, Post Keynesian economics, and system dynamics: Three strands of a heterodox economics braid* (pp. 1–37). In: 26 October 2004. https://www.researchgate.net/

Radzicki, M. J. (2021). System dynamics, data science, and institutional analysis. In R. Y. Cavana, B. Dangerfield, & O V. Pavlov et al. (Eds.), *Feedback economics economic: Economic modeling with system dynamics* (pp. 271–291). Springer. https://doi.org/10.1007/978-3-030-67190-7_19

Radzicki, M. J., & Tauheed, L. (2009). In defense of system dynamics: A response to professor hayden. *Journal of Economic Issues, 43*(4), 1043–1061. https://doi.org/10.2753/JEI0021-3624430411

Ramey, G., & Ramey, V. A. (1995). Cross-country evidence on the link between volatility and growth. *American Economic Review.*

Rodrik, D. (1999). Democracies pay higher wages. *Quarterly Journal of Economics.* https://doi.org/10.1162/003355399556115

Saviotti, P. P. (1996). *Technological evolution, variety, and the economy.* E. Elgar.

Sheng, L., & Yang, D. T. (2016). Expanding export variety: The role of institutional reforms in developing countries. *Journal of Development Economics, 118*, 45–58. Elsevier B.V. https://www.scopus.com/inward/record.uri?eid=2-s2.0-84944769092&doi=10.1016%2Fj.jdeveco.2015.08.009&partnerID=40&md5=9fa15bf7b6978f979e3fe124693ad8a6

Warhuus, J. P., & Basaiawmoit, R. V. (2014). Entrepreneurship education at Nordic technical higher education institutions: Comparing and contrasting program designs and content. *International Journal of Management Education.* https://doi.org/10.1016/j.ijme.2014.07.004

Winkler, C., Troudt, E., & Schweikert, C. et al. (2015). Infusing business and entrepreneurship education into a computer science curriculum—A case study of the stem virtual enterprise. *Journal of Business and Entrepreneurship.*

Zhu, S., & Fu, X. (2013). Drivers of export upgrading. *World Development, 51*, 221–233. Elsevier Ltd. https://www.scopus.com/inward/record.uri?eid=2-s2.0-84880674168&doi=10.1016%2Fj.worlddev.2013.05.017&partnerID=40&md5=64bf02fba23d33d9052a9ac1b2a0a62f

CHAPTER 3

Towards the Generation of a Green Technology Index

Alberto Méndez-Morales⊙

Abstract This chapter generates a proposal for a novel Green Technology Index. This index aims to identify if technological innovation systems behave well regarding the seven functions of those systems proposed in the literature, proposing a measure in which not only the development of green technologies is measured but a first approximation to a holistic view of what will be a well-functioning green technology innovation system. The proposed measure is essential given that global warming is a constant issue in the twenty-first century, and researchers believe that one meaningful way to solve this issue is by creating new green technologies that help relieve the pace of global warming. In addition, the proposed measure allows policymakers to enhance and redirect efforts to contribute to creating policies to help in what is, perhaps, the biggest challenge in our time.

A. Méndez-Morales (✉)
EGADE Business School, Tecnológico de Monterrey, México City, México
e-mail: amendez@tec.mx

© The Author(s), under exclusive license to Springer Nature
Switzerland AG 2023
M. M. Herrera (ed.), *Business Model Innovation for Energy Transition*,
Palgrave Studies in Democracy, Innovation, and Entrepreneurship
for Growth, https://doi.org/10.1007/978-3-031-34793-1_3

Keywords Green Technology Index · Technology Innovation System · Environmental Performance · Green Technologies · Innovation

3.1   INTRODUCTION

The use of novel green business models does not necessarily have to focus on the use of sustainable advanced technologies; however, the use of these types of technologies sure will strengthen the possibilities of those business models to succeed and to have positive effects on global warming issues globally. At the same time, the possibilities of new ventures to succeed in markets are deeply related to the technological framework in which the venture is developed and exploited. Therefore, in this chapter, we want to develop a measure related to the National Innovations Systems literature. Specifically, the functions of that innovation system, proposing a measure in which not only the development of green technologies is measured but a first approximation to a holistic view of what will be a well-functioning green technology innovation system.

As it seems evident in this case, the first question to answer is why we need a measure of green technologies innovation systems; the first evident answer is that we are running out of time to change the path to global warming. A measure like the one we are proposing will not only serve to rank regions or countries but will also determine if the public policies and private efforts are generating enough traction for stakeholders working around sustainable technologies. At the same time, it will help as a control barometer for understanding how green technologies have been developed and how the innovation system is working to create a better framework for new and old ventures trying to contribute to slowing the pace of global warming.

In this chapter, first, we want to show how a complex concept, such as the green technologies innovation system, can be measured. In this case, we use examples of other related indexes measuring complex phenomena such as entrepreneurship, innovation, or social development. Second, we review the functions of a Technological Innovation System; third, we propose a new index for green technology innovation systems to rank regions and countries based on Hekkert et al. (2007), Hekkert and Negro (2008) developments about the functions of a technological innovation system. Finally, we link the development of a measure, such as the one we propose, to developing green technology business models globally.

3.2 WHY USE INDEXES TO MEASURE COMPLEX CONCEPTS?

In the last decades, social researchers have tried to determine the factors creating the differences between developed and underdeveloped countries in several topics. As a result, different composite indicators, called indexes, have been created in recent years. The main idea behind those indexes is to create a measure in which different edges of a complex concept are joined to measure the performance of regions or countries. However, regarding indexes, a question arises, why use indexes instead of using individual indicators related to those concepts?

Most of the time, researchers must evaluate complex social phenomena that can not be described with singular statistics. For instance, if a researcher needs to measure the economic growth of a country, the per capita gross domestic product (GDP) can be used as an indicator of the activity of cities or regions; even with all the issues and critics of this famous imperfect indicator (Kalimeris et al., 2020), one should agree that it is an excellent single indicator of what is happening in the economy. The problem comes when a researcher tries to measure a country's social development or equality using the GDP because that individual indicator is not a good measure of that concept.

The researcher will face several problems given that social development is a complex phenomenon for which we, in theory, do not have a single accepted measure. In this case, the researcher may use several input measures to characterize the different aspects of this complex phenomenon, such as literacy, access to health services, transportation, life expectancy, access to water and sanitation services, and several others (Stern et al., 2022).

Thus, the appearance of indexes answers the necessity of researchers, policymakers, and society to measure complex societal challenges that will need the attention of those stakeholders but that cannot be measured easily. For researchers, indexes give them a better view of the differences between countries, which can guide them in understanding possible approaches and proposing possible solutions. In the case of policymakers, indexes give them a broader vision of the social phenomena to understand how the policies implemented in other regions, the index winners, can help them to develop new and improved policies and solutions to societal challenges but also can help them to prioritize efforts. Finally, in the case of society, indexes can help to attract attention to the differences between countries and create awareness of the issues suffered by different regions and countries in an easy way to communicate to the non-academic population (OECD, 2008).

Many indexes were created in recent years, and we do not intend to create an extensive inventory of them; however, we want to highlight some created to measure societal challenges.

- **Social Progress Index (SPI)**: Social Progress Index

    – The Social Progress Imperative programme has calculated the SPI since 2011. The main goal of the index is to measure social progress (a positive way to measure inequality) and to stimulate better social policies, especially for countries in which the index results are poor. In addition, the Social Progress Imperative reinforces that indicators, such as the per cápita GDP, are not suited to measure the progress of society.

- **Human Development Index (HDI)**: Human Development Reports

    – The HDI is a composite index calculated by the United Nations Development Programme (UNDP). The HDI is a member of the family of composite indexes of the United Nations; the Inequality-adjusted Human Development Index (IHDI), the Gender Development Index (GDI), the Multidimensional Poverty Index (MPI), and the Planetary pressures-adjusted Human Development Index (PH-DI). The HDI comprises per cápita income, life expectancy, and education indicators.

- **Corruption Perception Index (CPI)**: Transparency.org

    – In 1995, the non-governmental organization Transparency International launched the CPI to measure citizens' perceptions of corruption in the public sector. The initial methodology of the index was to merge several polls of corruption perception; however, in recent years, the methodology has also included other raw data to measure government policies developed to fight against corruption.

- **Environmental performance index (EPI)**: Environmental Performance Index

  - Since 2006, Yale University has measured the EPI to provide quantitative metrics for evaluating a country's environmental performance (Wolf et al., 2022). The EPI is divided into forty sustainability indicators, eleven categories, and three policy objectives and is designed to spot sustainability issues worldwide and as a tool to develop better policies to favour sustainability.

- **Global Competitiveness Report (GCR)**: Global Competitiveness Report 2020 | World Economic Forum (weforum.org)

  - Calculated since 2005 by the World Economic Forum (WEF), the GCR has as its primary objective to measure the interaction between Policies, institutions, and factors affecting a country's productivity. According to the WEF, the 2020 index comprises eleven priority pillars, twenty concept sub-pillars, and sixty-four individual indicators. Inside the individual indicators of the GCR it is possible to find some index measures like the Inclusive Internet Index, CPI, Egalitarian Democracy Index, E-Participation Index, and others.

As with the shown composite indicators, there are several topics in which researchers want to compare the results among countries and regions in order to improve government policies and create awareness over specific issues. For example, some composite indicators are also created to compare nations' development and public policies regarding science, technology, innovation, and entrepreneurship. Some of those indexes will be listed ahead:

- **Global Innovation Index (GII)**: Global Innovation Index 2022: What is the future of innovation-driven growth? (wipo.int)
  - GII is an index developed by INSEAD Business School and World Business Magazine in 2007 and has been calculated since 2021 by the World Intellectual Property Organization (WIPO). Since its beginnings, the index has been divided into two pillars denominated input or resources and outputs or results of innovation. For 2023, the index comprises 81 indicators measuring factors like institutions, human capital, infrastructure, market and business sophistication, knowledge and technology outputs, knowledge impact and diffusion, and creative outputs.

- **Global Entrepreneurship Monitor (GEM)**: GEM Global Entrepreneurship Monitor (gemconsortium.org)
  - GEM was introduced in 1999 in a joint effort among the London School of Business, Babson College, and Kauffman Centre. The index comprises secondary data, population surveys, and interviews with key players in the national entrepreneurship ecosystems. The main goal of the GEM is to identify the countries in which entrepreneurial activity is more robust, along with the generation of valuable data to analyse and design new policies to increase entrepreneurial activity in studied countries.

- **Quality Index for Latin-American Patents (QILAP)**: A novel quality index for Latin-American inventions - ScienceDirect
  - QILAP is an index developed by Méndez-Morales et al. (2022) for the 1996–2016 period. The main idea of the index is to compare the quality of patents instead of their quantity, given that not all patents are considered innovations but inventions.

The index comprises seven different measures of quality and is composed of eight different industry composite measures.

As can be seen, complex concepts like social progress, human development, corruption, environmental performance, competitiveness, inventiveness, innovation, and entrepreneurial activity can be measured using indexes. However, those concepts could hardly be measured using individual indicators, and therefore, the existence of those indexes contributes to the overarching definition, tackle analysis, and development of policies of the studied phenomena.

3.3 Green Technology Measures

At the same time, global warming has been raising the interests of the population, media, policymakers, and academia, and even when some political groups are neglecting it (Otteni & Weisskircher, 2022), there is a vast consensus about its dangerous consequences (Cruz et al., 2021). However, as discussed previously, global warming is a complex issue for which the possible solutions are not easy, and policies created to ease this phenomenon cannot be effortlessly evaluated.

In this regard, studies about the consequences and possible solutions to global warming are welcome. After searching in the Scopus database for research related to global warming and sustainable and green technologies, an increasing number of papers show the interest of academia in researching this topic, with a compound annual growth of 18.89% between 1990 and 2022 (Fig. 3.1). The used equation refers to the engineering field for the 1990–2022 period.

(global AND warming AND technologies) OR (green AND technologies) OR (sustainable AND technologies) AND (LIMIT-TO (SUBJAREA, "ENGI")) AND (LIMIT-TO (PUBYEAR,1990) OR LIMIT-TO (UBYEAR,2022)).

In 2022, more than 181 thousand green-related engineering papers were published in the Scopus database, against 714 in 1990; those numbers reflect the increasing interest of engineering academics to research topics related to the solution and adaptation to global warming phenomena and their related technologies. At the same time, China has the highest number of publications in the 1990–2022 period

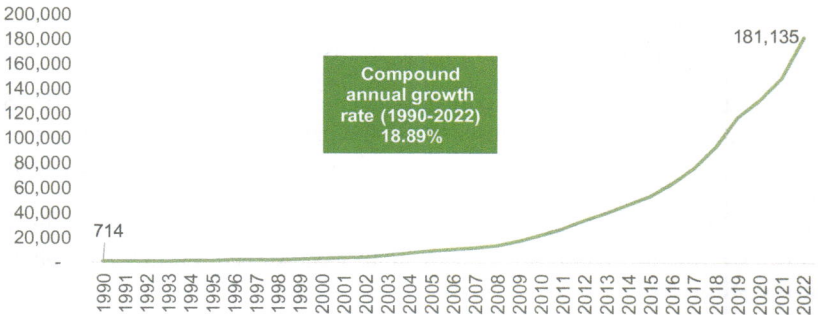

Fig. 3.1 Published articles on global warming, green and sustainable technologies in Scopus (1990–2022) (*Source* Scopus)

(Fig. 3.2), with almost 350 thousand documents. However, when population normalization is calculated, Australia seems to be the country in which the highest number of documents has been published, with 1542 papers per million inhabitants; this lets us know in which countries exist a more profound concern about global warming among academics and engineers.

At the same time, new technological development is needed to ease the effects of global warming. According to the economics of the innovation field, patents are an excellent way to measure the development of those technologies (Herrera-Ramírez et al., 2021; Méndez-Morales et al., 2022). The OECD has an adequate database in which the development of triadic (Japan, Europe, and the United States) environmental patents is measured. As shown in Fig. 3.3, for 2019, China had a higher number of environmental family patents, followed closely by the United States and Japan.

The joint information in Figs. 3.2 and 3.3 shows that China is a preponderant country in research and development of green environmental technologies. Also, The United States seems to be an important country in this logic. However, to combat global warming, it is essential to create knowledge and technology and implement both to diminish the environmental damage. A partial way to understand if this technology is implemented is to see the proportion of renewable energy used in a determined country.

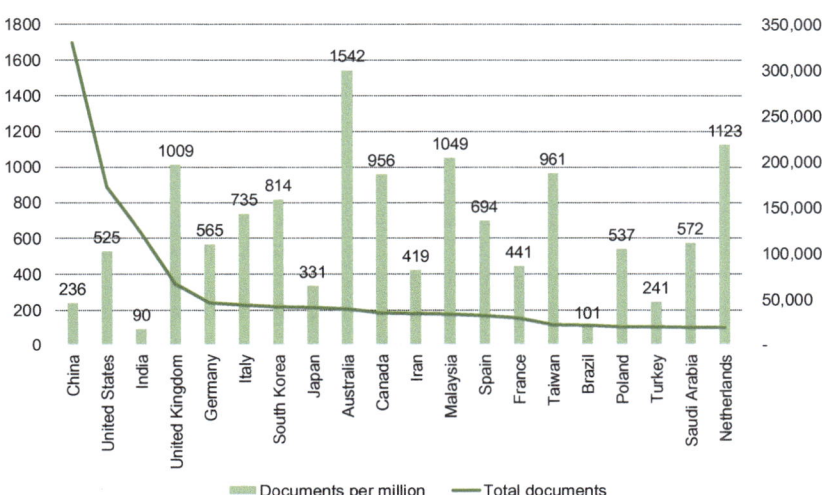

Fig. 3.2 Green technologies publications in Scopus for the period 1990–2022 (*Source* Scopus)

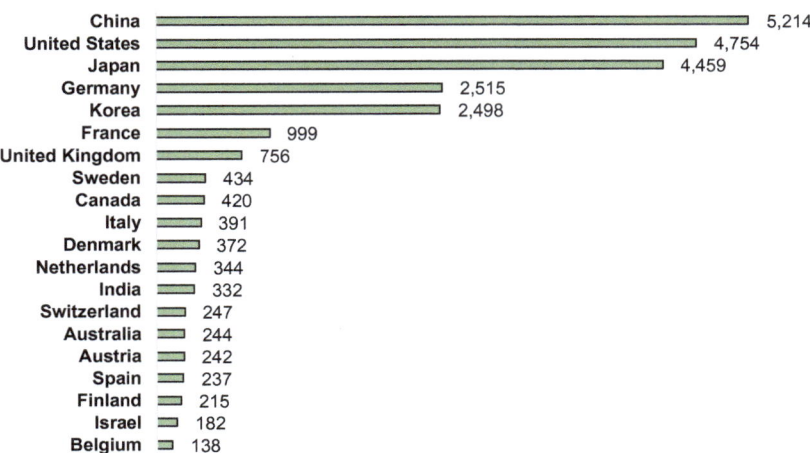

Fig. 3.3 Total environmental triadic family patents per country (2019) (*Source* OECD.Stat)

As seen in Fig. 3.4, the countries in which renewable energies present a higher proportion of the total energy grid are not the same, for which there are a higher number of papers and patents. Countries like Iceland, Paraguay, and Laos use geothermal energy and hydropower to cover their internal energy necessities, creating more energy than the country needs but are not winners in developing green knowledge and technology. Conversely, even when academic and technological development is excellent for countries such as China, The United States, or Japan, that does not mean that implementing green energy technologies is easy for these countries. A country must create and implement green technologies simultaneously, which is impossible in all countries. Paraguay, for example, can use rivers to create energy, but it will depend on external technology to exploit it, given that it has poor results related to academic papers or patents. China, in the other hand, can create thousands of papers and patents, however, this does not mean that is implementing that knowledge and technology in its territory. These examples show that measuring green technologies' development and implementation is not easy; therefore, combining several indicators to understand this complex phenomenon is necessary.

In Fig. 3.5, the combination of papers, patents, and renewable energy grid use are merged; in this case, we are measuring the countries in which intellectual capacities (papers), technology development (papers),

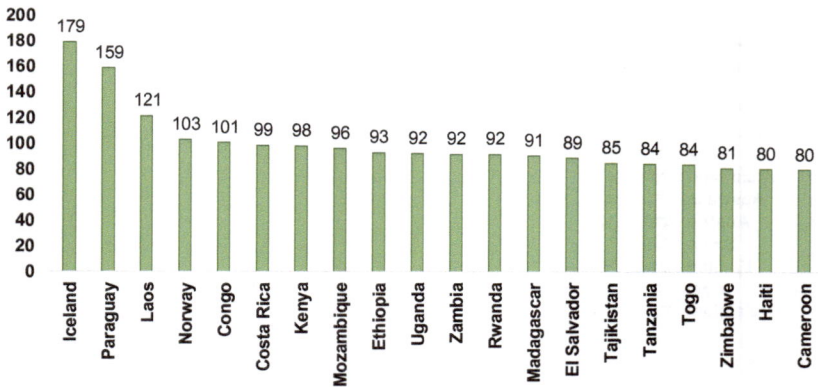

Fig. 3.4 Proportion of renewable energy over total energy production in 2020 (*Source* OECD.Stat)

and implementation (renewable energy grid) are robust. For example, Sweden has the intellectual capacity to publish 1600 green technology papers per million inhabitants, the inventive capacity to generate 43 green patents per million inhabitants, and the use of 73.6% of renewable energies in its entire energy grid. On the other hand, Korea creates many patents (48 per million inhabitants) but less knowledge than Sweden (814 papers per million) and is behind in implementing those energy technologies, with only 3.6% of the energy grid supported by renewable energies. At the same time, in Figs. 3.2 and 3.3, it seems that China has the optimal position related to papers and patents production; however, those quantities are inferior when measures are normalized, and renewable energy use (16.6%) is poor compared to countries such as Sweden (73.6%), Germany (28.1%), or Italy (33%).

Is this moment worth asking what a country like Sweden is doing to have the possibility to create knowledge, develop technologies, and implement them in energy production at the same time? Even when the answer is not easy, the theory behind national, and technological innovation systems can help to understand the differences among those countries.

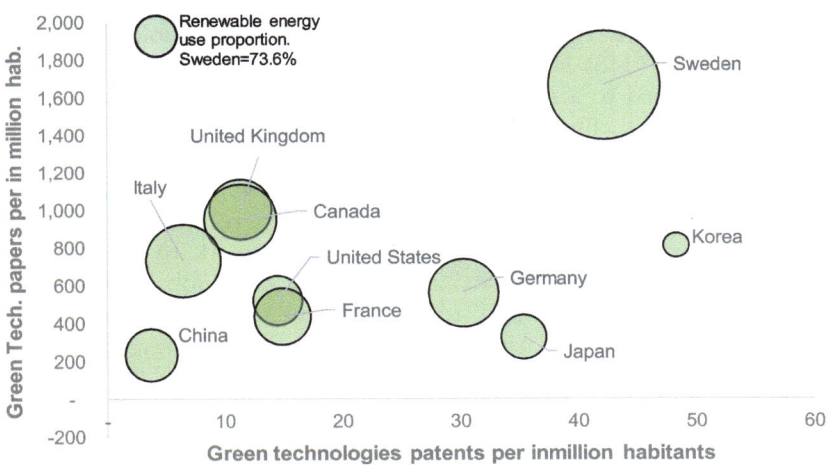

Fig. 3.5 Intellectual and technology capacities vs Renewable energies grid in selected countries (*Source* OECD.Stat. & Scopus)

3.4   NATIONAL AND TECHNOLOGICAL INNOVATION SYSTEMS

In the developments of Freeman (1987, 1995), Lundvall (1999, 2010), and Nelson (1993), the National Innovation Systems (NIS) arose as an institutional, cultural, and political framework that contributes to the development and diffusion of new technologies under the concept of innovation. In the case of Sweden, it is possible to assume that the framework in which green technologies arise facilitates the development and implementation of those technologies. Institutions such as universities, markets, governments, and companies are aligned to create knowledge and develop and implement those technologies.

When discussing the function of green technologies in reducing the effects of global warming, it seems logical to think that an institutional framework is needed to develop those technologies and reduce global warming; it is possible that in the case of Sweden, that framework is aligned in terms of policies, private markets, academia, and society, which allows the country to obtain better results in terms of green technology development and implementation.

However, the concept of NIS seems too broad regarding a specific technology market. In this case, the definition of Technological Innovation Systems (TIS) is focused on a specific market for technologies but circumscribed to a specific region. Carlsson and Stankiewicz (1991, p. 111) defined technological innovation systems as "... *a network of agents interacting in a specific economic/industrial area under a particular institutional framework ... involved in the generation, diffusion and utilization of technology.*" In this case, the green technology system is related to the capacity of the institutional framework to develop and implement these technological innovations to ease global warming in a specific country. According to only a few indicators proposed in Fig. 3.5, Sweden could have a better Green Technology Innovation System (GTIS).

Nevertheless, if a country wants to reach or surpass the development of the GTIS of Sweden to ease global warming in its territory, it needs to understand that the concept of TIS is complex as the ones reviewed in previous sections. Thus, we believe an index must be developed to measure the GTIS framework.

According to the works of Hekkert (Hekkert & Negro, 2008; Hekkert et al., 2007), the TIS has seven different functions. Therefore, a country must fulfil these functions to have a well-functioning system.

- *Function 1: Entrepreneurial activities.*

  – Entrepreneurs are the heart of any innovation system. Without the presence of entrepreneurs, developing and implementing technologies is impossible. The primary role of entrepreneurs is to profit from the technology market by creating, transferring, and exploiting technologies in uncertain markets. Then, a healthy TIS needs to allow the creation and development of new ventures.

- *Function 2: Knowledge development.*

  – In order to create innovations embedded into technologies, it is necessary to create the basic knowledge supporting technology development. According to Hekkert et al. (2007), this knowledge can be gathered by learning by doing and learning by searching. The process of learning by doing is most common for entrepreneurs and mature firms, and the process of learning by searching is most common in universities and research centres. Therefore, searching for knowledge development data in TIS is possible depending on the technology maturity level.

- *Function 3: Knowledge diffusion.*

  – Knowledge creation is not enough to generate a healthy TIS; this knowledge needs to be transferred to other markets and society in several ways. First, the market needs rules to arrange and use this knowledge once transferred. Second, knowledge needs to be available to the majority of the market, and third, knowledge creators need to obtain profitability for their tasks.

- *Function 4: Guidance of the search.*
  - Given that no unlimited funds exist for Research and Development (R&D), the system needs institutional guidance on where to put those resources. Governments, leading companies, research institutions, clients, and competitors can guide the knowledge creation and exploitation of companies and research institutions in technology markets.

- *Function 5: Market Formation.*
  - The diffusion and use of technologies are necessary for the TIS to operate appropriately. Thus, measuring the deep and scope of that technological diffusion is necessary. To the extent that the technology is used, adapted, made suitable, and transformed, we can assume there is a market for those technologies.

- *Function 6: Resource mobilization.*
  - R&D, knowledge diffusion, and market formation are costly tasks, and given the uncertainty of technology development, funds to support those tasks could be scarce, especially when markets are not fully developed. Thus, the company's internal funds, financial markets, government subsidies, venture capital, and other sources of funds are necessary to create a functioning TIS.

- *Function 7: Creation of legitimacy.*
  - According to Suchman (1995, p. 574), legitimacy is "a generalized perception that the actions of an entity are desirable, proper, or appropriate within some socially constructed system of norms, values, beliefs, and definitions." In this case, that system is the TIS. Function 7 describes the interrelations in the

technology market that show that specific technology adoption is desirable.

Overall, a well-performing TIS will be one in which all these featured functions are fulfiled and interconnected. For example, in the case of Sweden, one can hypothesize that there are connections between the creation of knowledge as papers, the diffusion of knowledge in the form of triadic patent applications, and the successful implementation of renewable energy sources in its grid; thus, the general behaviour of green TIS is better than the one of Korea, China, and other studied countries in Fig. 3.5. In the next section, we will propose the development of a composed index to measure the development of Green TIS based on the seven functions of the system.

Generally, the well-performing NIS and TIS concept is rooted in Carayannis et al. (2012) quintuple helix innovation model. In the case of Sweden, one can think that the creation and enhancement of an ecosystem in which companies, universities, government (triple helix), and society (quadruple helix) are put over the table to create an innovation system in which the challenge of global warming is the focus (quintuple helix), allows the country to behave better than other countries to protect the environment.

3.5   Proposed Index

As it was reviewed in the first part of the chapter, measuring a complex phenomenon like the development of a TIS requires to create also a complex composed measure, the Index of Green Technologies Innovation System (IGTIS).

Based on the developments of Hekkert et al. (2007), IGTIS is composed of seven pillars, each of one representing one of the functions of the TIS. Figure 3.6 presents an overview of the IGTIS structure.

The equation representing the IGTIS will be:

$$IGTIS_{c,y} = w_1 * EA_{c,y} + w_2 * KD_{c,y} + w_3 * Dif_{c,y} + w_4 * GS_{c,y}$$
$$+ w_5 * MF_{c,y} + w_6 * Res_{c,y} + w_7 * Leg_{c,y}$$

where:
   EA: Entrepreneurial activities.
   KD: Knowledge Development.
   Dif: Knowledge Diffusion.

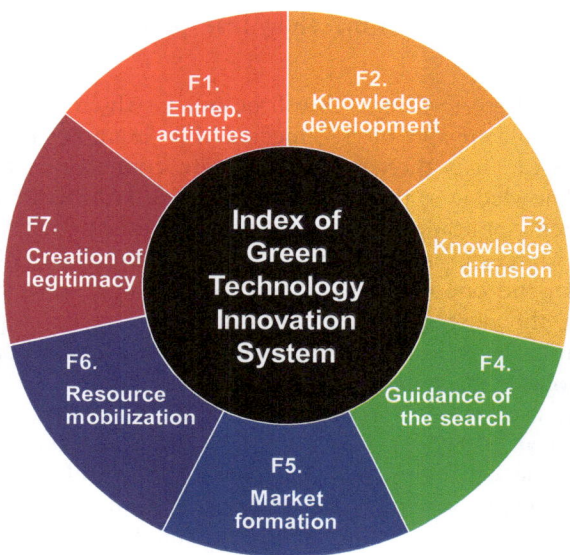

Fig. 3.6 Functions of the green technology innovation system (*Source* Hekkert [2007])

GS: Guidance of the Search.
MF: Market Formation.
Res: Resource Mobilization.

Leg: Creation of Legitimacy.

Those measures represent each of the TIS functions proposed by Hekkert. At the same time, w_1 to w_7 represent the function weights. Given that when creating a composed index, there is always an intense debate about the weights, in this case, the first approach will be to follow the methodology of Méndez-Morales et al. (2021, 2022), and therefore the seven pillars will have an equal weight of one-seventh, thus $w_1 = w_2 = w_3 = w_4 = w_5 = w_6 = w_7$. In our proposal, the IGTIS's pillars will be decomposed into twenty-six variables composing each TIS function. Therefore, the second approach will be that each composing measure weighs one twenty-sixth; that implies that the TIS functions with a higher number of composing variables will be more important than others in terms of weights. In Appendix 1, a resume of the two approaches is presented.

Ahead, we present an overview of the proposed indicator composing the IGTI. Then, in brackets, we present a proposed data source to gather this information. Each time we try to use free or low-cost databases to gather this information; however, most of the variables in function 6, i.e., resource mobilization, need specialized financial data.

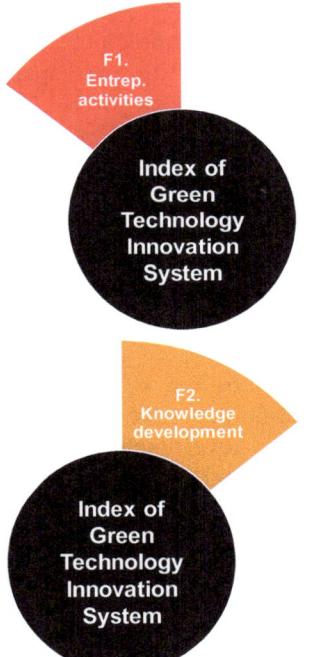

EA.1. Creation of entrepreneurship in green technologies (GEM).

EA.2. Entrepreneurial framework conditions (GEM).

EA.3. The proportion of green technology companies in stock markets (National stock market).

EA.4. The proportion of market capitalization of green technology companies (National stock markets).

KD.1. Quantity of green patent families (WIPO or National Patent offices).

KD.2. Green patent families per million inhabitants (WIPO, National Patent offices, OCDE).

KD.3. STEM Graduates per capita (UNESCO UIS).

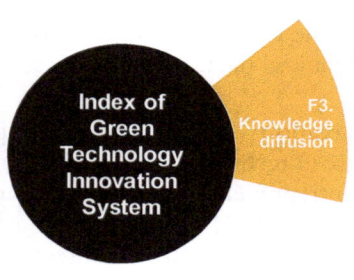

Dif.1.   The number of research centres specializing in green technologies (National Bureau of Research).

Dif.2.   Green technologies scientific papers per capita (Scopus, Web of Science).

Dif.3.   Green technologies scientific papers whit the participation of international entities (Scopus, Web of Science).

Dif.4.   The number of scientific journals specializing in green technology (SCIMAGO).

Dif.5.   Average citation in green technology scientific papers (Scopus, Web of Science).

GS.1.   The number of specific green technology laws signed by the government (Countries' Congress).

GS.2.   The number of times the past economic and social development plans used words related to global warming or green technologies (National development bureaus).

GS.3.   International signed agreements for environmental conservation (Multilateral institutions such as UNO or CEPAL).

MF.1.   Proportion of population with access to clean technologies to cook (UNESCO, OECD).

MF.2.   Population with access to green energy (UNESCO, UNO, OECD).

MF.3.   C02 emissions generated by electricity production (national energy authorities).

MF.4.   Energy efficiency score (RISE).

Res.1.   Green Technology transfer expenditure by country (WIPO).

Res.2.   Venture capital investments in green technology firms (Refinitiv, Bloomberg).

Res.3.   Per GDP green bonds issued by country (Refinitiv, Bloomberg).

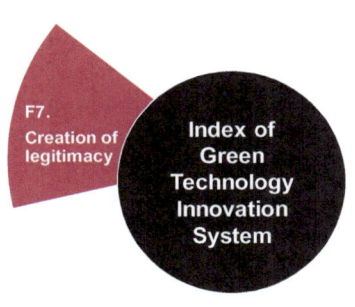

Leg.1.   Installed capacity for Energy generation using green technologies (UNO, UNESCO, OECD).

Leg.2.   Public expenditure in green energies as a percentage of GDP (IDB).

Leg.3.   Per GDP value of green energies government auctions (Countries' Energy Bureaus).

Leg.4.   Industry greenhouse emissions by country (World Population Review).

After gathering the data for selected nations, it is necessary to normalize this data. For example, some variables are calculated as a percentage of GDP, like Leg.3 or Res.3., while others are calculated as raw numbers, like Dif.1. or KD.1.; thus, this data needs to be calculated as normalized indicators, for what we recommend to follow the methodology of Méndez-Morales et al. (2021, 2022). Therefore, the value of the normalized variables.

will be:

$$\text{Normalized Variable}_{c,t} = \frac{\log_{10}\left(\text{Var.i}_{c,t} + 1\right) - \log_{10}\min((\text{Var.i}_t + 1))}{\log_{10}\max((\text{Var.i}_t + 1)) - \log_{10}\min((\text{Var.i}_t + 1))}$$

Typically, country measures will follow a log-normal distribution; thus, it is necessary to calculate the logarithms of each variable before normalizing the indicators. In this case, we propose a base ten logarithm; however, another base could be selected according to the type of distribution for each variable. In the equation, c and t represent country and time measures; thus, we propose a min–max normalization for each country and time point in which the maximum value of each variable will be one and the minimum zero. Finally, with the two approaches, the maximum value for a given c in any given t will be one hundred, and the minimum will be zero. In terms of operativity, the information for the index could be gathered using country allies representing universities, research centres, or government organizations.

3.6   THE GREEN BUSINESS MODELS AND IGTIS

When discussing the development of new technologies, it seems evident that those new technologies represent innovation. At the same time, that innovation could be created in established or new companies. Nevertheless, according to Schumpeter's Mark I theory, this development is developed mainly by newcomers. The startup and unicorn boom of recent years, fueled by venture capital investments, reveals that those new entrepreneurs seem to be more prepared to develop new green technologies. Nowadays, innovation and technology can no longer be separated from green conservation, not only because technology can help diminish global warming but because innovative business models need to reinforce their social impact by implementing strategies to diminish their impact.

The proposed index can help the development of green business models in various ways. First, the measure and comparison of a green technology index will help policymakers to determine the future policies to implement to increase the number of green entrepreneurs that, at the same time, will generate positive spillovers in the TIS. Governments and universities need to model the green technological system and evaluate public policies in the long term for specific technologies like solar and wind energy generation.

At the same time, the measurement results will be a signal for entrepreneurs needing to understand which localization will be favourable for locating their new companies. Fourth, universities would understand if the efforts related to green technology generation create enough traction to accomplish green technology transfer to new and established firms. Fifth, society will understand if the country's institutionalist is solving one of the most challenging issues of our times.

Appendix 1: Possible weight approaches to calculate IGTIS

Approach 1. Same Weight for each variable of the TIS						
F1. Entrepreneurial activities.	**F2. Knowledge development.**	**F3. Knowledge diffusion.**	**F4. Guidance of the search**	**F5. Market Formation.**	**F6. Resource mobilization.**	**F7. Creation of legitimacy.**
Function weight=0,143	Function weight=0,143	Function weight=0,143	Function weight=0,143	Function weight=0,143	Function weight=0,143	Function weight=0,143
EA.1. Creation of entrepreneurship in green technologies. Weight=0,036.	KD.1. Quantity of green patent families. Weight=0,048.	Dif.1. The number of research centers specializing in green technologies. Weight=0,029.	GS.1. The number of specific green technology laws signed by the government. Weight=0,048.	MF.1. Number Proportion of population with access to clean technologies to cook. Weight=0,036.	Res.1. Green Technology transfer expenditure by country. Weight=0,048.	Leg.1. Installed capacity for Energy generation using green technologies. Weight=0,036.
EA.2. Entrepreneurial framework conditions. Weight=0,036.	KD.2. Green patent families per million inhabitants. Weight=0,048.	Dif.2. Green technologies scientific papers per capita. Weight=0,029.	GS.2. The number of times the past economic and social development plans used words related to global warming or green technologies. Weight=0,048.	MF.2. Population with access to green energy. Weight=0,36.	Res.2. Venture capital investments in green technology firms. Weight=0,048.	Leg.2. Public expenditure in green energies as a percentage of GDP. Weight=0,036.
EA.3. The proportion of green technology companies in stock markets. Weight=0,036.	KD.3. STEM Graduates per capita. Weight=0,048.	Dif.3. Green technologies scientific papers whit the participation of international entities. Weight=0,029.	GS.3. International signed agreements for environmental conservation. Weight=0,048.	MF.3. C02 emissions generated by electricity production. Weight=0,036.	Res.3. Per GDP Green bonds issued by country. Weight=0,048.	Leg.3. Per GDP value of green energies government auctions. Weight=0,036.
EA.4. The proportion of market capitalization of green technology companies. Weight=0,036.		Dif.4. The number of scientific journals specializing in green technology. Weight=0,029.		MF.4. Energy efficiency score. Weight=0,036.		Leg.4. Industry greenhouse emissions by country. Weight=0,036.
		Dif.5. Average citation in green technology scientific papers. Weight=0,029.				

Approach 2. Same Weight for each variable of the TIS						
F1. Entrepreneurial activities.	F2. Knowledge development.	F3. Knowledge diffusion.	F4. Guidance of the search	F5. Market Formation.	F6. Resource mobilization.	F7. Creation of legitimacy.
Function weight=0,154	Function weight=0,115	Function weight=0,192	Function weight=0,115	Function weight=0,154	Function weight=0,115	Function weight=0,154
EA.1. Creation of entrepreneurship in green technologies. Weight=0,038.	KD.1. Quantity of green patent families. Weight=0,038.	Dif.1. The number of research centers specializing in green technologies. Weight=0,038.	GS.1. The number of specific green technology laws signed by the government. Weight=0,038.	MF.1. Number Proportion of population with access to clean technologies to cook. Weight=0,038.	Res.1. Green Technology transfer expenditure by country. Weight=0,038.	Leg.1. Installed capacity for Energy generation using green technologies. Weight=0,038.
EA.2. Entrepreneurial framework conditions. Weight=0,038.	KD.2. Green patent families per million inhabitants. Weight=0,038.	Dif.2. Green technologies scientific papers per capita. Weight=0,038.	GS.2. The number of times the past economic and social development plans used words related to global warming or green technologies. Weight=0,038.	MF.2. Population with access to green energy. Weight=0,38.	Res.2. Venture capital investments in green technology firms. Weight=0,038.	Leg.2. Public expenditure in green energies as a percentage of GDP. Weight=0,038.
EA.3. The proportion of green technology companies in stock markets. Weight=0,038.	KD.3. STEM Graduates per capita. Weight=0,038.	Dif.3. Green technologies scientific papers whit the participation of international entities. Weight=0,038.	GS.3. International signed agreements for environmental conservation. Weight=0,038.	MF.3. C02 emissions generated by electricity production. Weight=0,038.	Res.3. Per GDP Green bonds issued by country. Weight=0,038.	Leg.3. Per GDP value of green energies government auctions. Weight=0,038.
EA.4. The proportion of market capitalization of green technology companies. Weight=0,038.		Dif.4. The number of scientific journals specializing in green technology. Weight=0,038.		MF.4. Energy efficiency score. Weight=0,038.		Leg.4. Industry greenhouse emissions by country. Weight=0,038.
		Dif.5. Average citation in green technology scientific papers. Weight=0,038.				

REFERENCES

Carayannis, E. G., Barth, T. D., & Campbell, D. F. (2012). The Quintuple Helix innovation model: Global warming as a challenge and driver for innovation. *Journal of Innovation and Entrepreneurship, 1*(1), 2. https://doi.org/10.1186/2192-5372-1-2

Carlsson, B., & Stankiewicz, R. (1991). On the nature, function and composition of technological systems. *Journal of Evolutionary Economics, 1*(2), 93–118. https://doi.org/10.1007/BF01224915

Cruz, J. L., Rossi-Hansberg, E., & Griffin, K. C. (2021). *The economic geography of global warming.* https://doi.org/10.3386/W28466

Freeman, C. (1987). *Technology policy and economic performance: Lessons from Japan* (p. 34). Pinter. https://doi.org/10.1016/0048-7333(88)90011-X

Freeman, C. (1995). The "National System of Innovation" in historical perspective. In *Cambridge Journal of Economics, 19.*

Hekkert, M. P., & Negro, S. O. (2008). Functions of innovation systems as a
framework to understand sustainable technological change: Empirical evidence
for earlier claims. *Technological Forecasting & Social Change, 76,* 584–594.
https://doi.org/10.1016/j.techfore.2008.04.013

Hekkert, M. P., Suurs, R. A. A., Negro, S. O., Kuhlmann, S., & Smits, R.
E. H. M. (2007). Functions of innovation systems: A new approach for
analysing technological change. *Technological Forecasting and Social Change,
74*(4), 413–432. https://doi.org/10.1016/j.techfore.2006.03.002

Herrera-Ramírez, M. M., Méndez-Morales, A., & Barrios-Campos, D. M.
(2021). How the quality of patents contributes to revitalising the innovation
system?, Evidence from the renewable energy sector. *Revista de Investigación,
Desarrollo e Innovación, 11*(2), 227–242. https://doi.org/10.19053/202
78306.v11.n2.2021.12753

Kalimeris, P., Bithas, K., Richardson, C., & Nijkamp, P. (2020). Hidden linkages
between resources and economy: A "Beyond-GDP" approach using alternative welfare indicators. *Ecological Economics, 169,* 106508. https://doi.org/
10.1016/j.ecolecon.2019.106508

Lundvall, B.-Å. (1999). National business systems and national systems of innovation. *International Studies of Management & Organization, 29*(2), 60–77.
https://doi.org/10.1080/00208825.1999.11656763

Lundvall, B.-Å. (2010). National systems of Innovation: Toward a theory of
innovation and interactive learning. In *National systems of innovation: Toward
a theory of innovation and interactive learning.* Anthem. https://doi.org/10.
7135/UPO9781843318903

Méndez-Morales, A., Cuellar, S., Herrera, M. M., & Mejía, J. (2022). A novel
quality index for Latin-American inventions. *World Patent Information, 71,*
102154. https://doi.org/10.1016/J.WPI.2022.102154

Méndez-Morales, A., Ochoa-Urrego, R., & Randhir, T. O. (2021). Measuring
the quality of patents among Latin-American universities. *Studies in Higher
Education.* https://doi.org/10.1080/03075079.2021.2020749

Nelson, R. R. (1993). *National systems of innovation: A comparative analysis.*

OECD. (2008). Handbook on constructing composite indicators: Methodology
and user guide. In *Handbook on constructing composite indicators: Methodology
and user guide.* https://doi.org/10.1787/9789264043466-en

Otteni, C., & Weisskircher, M. (2022). Global warming and polarization: Wind
turbines and the electoral success of the greens and the populist radical right.
European Journal of Political Research, 61, 1102–1122. https://doi.org/10.
1111/1475-6765.12487

Stern, S., Harmacek, J., Krylova, P., & Htltch, M. (2022). *2022 Social progress
index.* https://www.socialprogress.org/

Suchman, M. C. (1995). Managing legitimacy: Strategic and institutional approaches. *Academy of Management Review, 20*(3), 571–610. https://doi.org/10.5465/amr.1995.9508080331

Wolf, M. J., Emerson, J. W., Esty, D. C., de Sherbinin, A., & Wendling, Z. A. (2022). *2022 Environmental performance index.* epi.yale.edu

Assessing Renewable Energy Projects: A Sustainability and Innovation Perspective

Javier Andres Calderon-Tellez and Milton M. Herrera◉

Abstract From past we learnt that all projects are not environmentally friendly at long-term range. In order to take environment into account, this paper seeks to establish a balance of innovation and sustainability for project management (PM) through a simulation model. The research uses actual data from a Colombian energy project, a developing country which will increase investments in solar projects for 2026. The methodology for simulation is based on system dynamics technique that includes three scenarios of process innovation. The paper offers a key element

J. A. Calderon-Tellez
Science Policy Research Unit (SPRU), University of Sussex Business School, University of Sussex, Brighton, UK

Ejército Nacional, Bogota, Colombia

M. M. Herrera (✉)
Economic Sciences Research Centre, Universidad Militar Nueva Granada, Bogotá, Colombia
e-mail: milton.herrera@unimilitar.edu.co; jc676@sussex.ac.uk

© The Author(s), under exclusive license to Springer Nature Switzerland AG 2023
M. M. Herrera (ed.), *Business Model Innovation for Energy Transition*, Palgrave Studies in Democracy, Innovation, and Entrepreneurship for Growth, https://doi.org/10.1007/978-3-031-34793-1_4

61

that changes project span within social, environmental, economic, and administrative sustainability aspects over time.

Keywords Process innovation · Management of Projects (MoP) · Project Management · Sustainability · System Dynamics · Simulation · Electricity Market

4.1   INTRODUCTION

The United Nations established the thirteenth sustainable development goal (SDG 13) about climate action. The SDG 13 pledges to limit warming to 1.5 °C global net carbon dioxide (CO_2) emissions. The SDG report shows that developed countries had decreased by 6.5% greenhouse gas emissions from 2000 to 2018. By contrast, developing countries had increased 43.2% greenhouse gas emissions from 2000 to 2013 (United Nations, 2020). It is urged to keep attention of greenhouse gas emissions in developing countries.

A potential solution to offset greenhouse gas emissions is the use of solar energy in developing countries. Coupled with the need to use solar energy, Colombia, a developing country, has planned 409 projects to generate 25,385 MW from 2021 to 2026, which represents 48.63% of total energy projected (UPME, 2021). However, developing projects of solar energy has innovation and sustainability aspects that must be studied at long term. This study is based on the impact produced within a solar energy project by enhancing process for innovation, and by social, environmental, economic, and administrative aspects for sustainability.

Adding innovation and sustainability to project management (PM) increase complexity to the project. In view of complexity, PM could be guided by a holistic approach (Elia et al., 2020; Silvius & de Graaf, 2019) handled by system dynamics (SD) modelling. On one hand, innovation has been driven through PM which projects have being radical (radical innovation) as was the Manhattan project (Lenfle et al., 2016) or changing the development of process (process innovation) (Mishra & Browning, 2020) as the Apollo project did. On the other hand, sustainability has been addressed through PM by project managers to complete the project (Silvius & de Graaf, 2019). However, impact caused after back-end of project is not taken in consideration to evaluate the sustainability consequences over time.

This chapter will help project managers and policymakers having a tool to forecast the consequences of developing a project at long range term. In addition, this research is limited for developing countries to help climate SDG 13 by solar energy to satisfy access to electricity SDG 7. To reach this investigation related to project's impact over time, this research raises the question: What are the effects on social, environmental, economic, and administrative sustainability aspects by increasing process innovation of an energy project?

The SD model proposed in this research will help for decision-making but the model won't make the decision if project should be developed or not. The simulation is able to forecast the sustainability behaviour over time. This research changes the development of process innovation by three scenarios. The results found the relationship between process innovation and project span. The innovative contribution in PM discipline is the extension of project life cycle through the simulation of project's impact after the end of its back-end.

The remainder of the chapter is structured as follows. The next section presents the theoretical background for innovation and PM, sustainability and PM, and SD and PM. It is followed by an outline of the methodology used for an SD approach which includes model description and model validation. Later, results section is structured into five main topics: (a) scenario design, (b) social sustainability dimension, (c) environmental sustainability dimension, (d) economic sustainability dimension, and (e) administrative with ethics sustainability dimension. Finally, the paper discusses the results, and then concludes and offers future research directions.

4.2 Theoretical Background

Project management has been focused on the triple constraint, however, this management limits to project execution or delivery. The management of projects has been treated as complex (Davies, 2019) and integrating innovation and sustainability to projects increases its complexity. However, system dynamics (SD) supports project complexity (Elia et al., 2020; Sterman, 2000). The literature review of PM has been concentrated in studies of innovation and PM, sustainability and PM, and SD and PM, as follows:

4.2.1 Innovation and Project Management

Innovation and PM has been interrelated since 1940s in military projects sponsored by the government where technological innovation took place. However, technology in military projects brings radical changes in processes which uncertainty level increased to complete it. A sciento-metrics analysis was developed to identify principal authors in innovation and PM, as shown in Fig. 4.1. The analysis converges that innovation in PM focuses on product and process innovation (Maniak & Midler, 2014; Mishra & Browning, 2020), radical innovation (Lenfle et al., 2016; Pich et al., 2002). In addition, innovation depends on knowledge sharing (Müller et al., 2013), endeavours between industry, university, and R&D collaborations (Fernandes et al., 2020; Lippe & Vom Brocke, 2016), and capability to share resources management (Bjorvatn & Wald, 2018; Matinheikki et al., 2016).

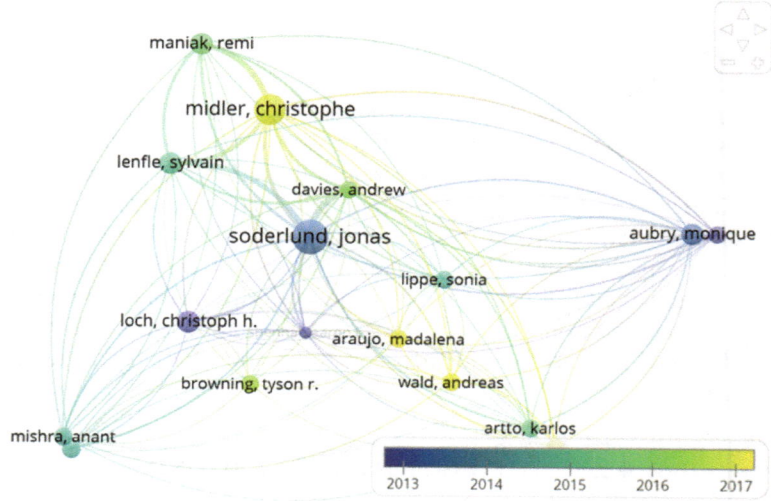

Fig. 4.1 Innovation and project management scientometrics analysis with the assistance of VOSviewer[1] based on map of authors (unit of analysis) and bibliographic coupling (type of analysis)

---

[1] VOSviewer is software designed to analyse and visualize bibliometric networks. It can create maps based on network data, bibliographic data, and text data using co-authorship, co-occurrence, citation, bibliographic coupling, or co-citation as types of analysis (van Eck & Waltman, 2010).

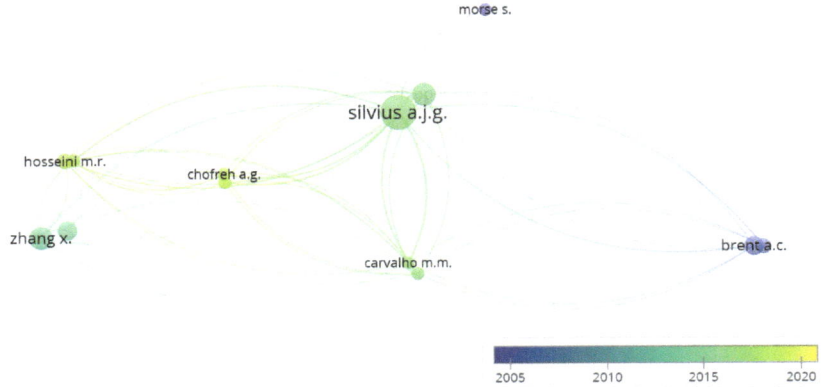

Fig. 4.2 Sustainability and project management scientometrics analysis with the assistance of VOSviewer based on map of authors (unit of analysis) and bibliographic coupling (type of analysis)

4.2.2 Sustainability and Project Management

There have been multiple studies that discussed sustainability and sustainable PM (Carvalho & Rabechini, 2017; Martens & Carvalho, 2016; Silvius & de Graaf, 2019). Studies of Carvalho (Carvalho & Rabechini, 2017), Martens (Martens & Carvalho, 2017), and Silvius (Silvius & de Graaf, 2019) relate sustainability to the triple bottom line (TBL)—social, environmental, and economic perspectives. A scientometrics analysis was carried out to determine directions and relationships between authors for sustainability in PM, as shown in Fig. 4.2. Several authors of sustainability in PM conclude that the sustainability is considered as complex, required as paradigm shift, and needed as holistic instead of systematic approach (Silvius & de Graaf, 2019).

4.2.3 System Dynamics and Project Management

System dynamics (SD) is a technique developed by Forrester (1961) to simulate delays and nonlinear dynamic behaviour of a complex system over time (Ansari, 2019; Sterman, 2000). SD has been applied in PM (Poziomek et al., 1977; Roberts, 1964) for decision-making from a systemic perspective (Chronéer & Backlund, 2015; Lopes et al., 2015;

Sadabadi & Kama, 2014). One of the most used SD models for PM is the rework cycle (Cooper, 1980), its structure is explained in model description section. The literature review conducted a scientometrics analysis, as shown in Fig. 4.3. The analysis identified SD models within PM discipline that include the triple constraints (Nasirzadeh et al., 2008), innovation (Lane & Husemann, 2008; Pargar et al., 2019; Park et al., 2004; van Oorschot et al., 2018), sustainability (Alasad & Motawa, 2015; Shen et al., 2005). However, there is a gap in linking innovation and sustainability with PM all together.

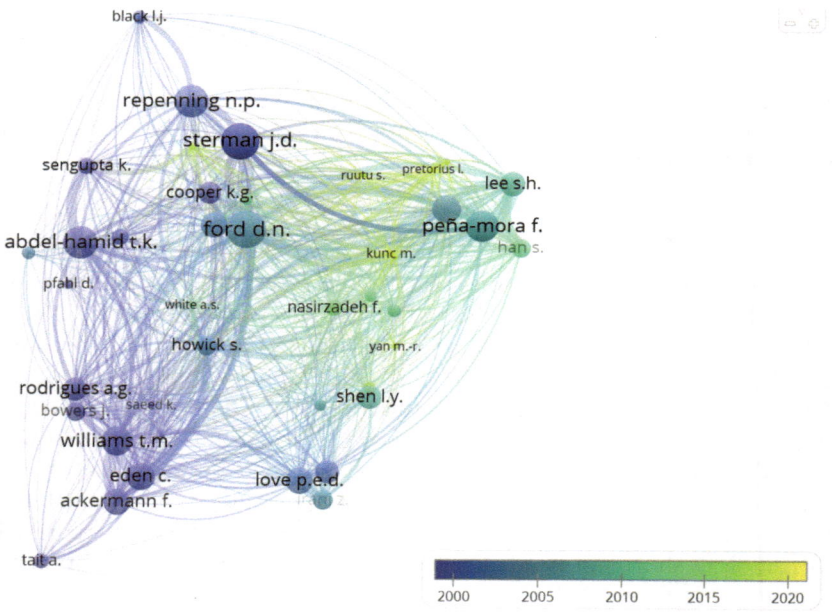

Fig. 4.3 System Dynamics and project management scientometrics analysis with the assistance of VOSviewer based on map of authors (unit of analysis) and bibliographic coupling (type of analysis)

4.3 METHODOLOGY

This research was conducted on SD methodology proposed by Sterman (2000). The hard systemic approach suggests interactive steps: problem articulation, dynamics hypothesis, formulation, testing, and policy formulation and evaluation. The first step, problem articulation, is defined as the implications and consequences of an energy project by using solar panels. The next steps are dynamic hypothesis and formulation which are explained in model description section by using a causal loop diagram. The fourth step is testing which is justified in model validation section. The last step, policy formulation and evaluation, are the results formulated from the three scenarios concerning to process innovation proposed in this research.

4.3.1 Model Description

An SD model for PM was built to understand the dynamics behaviour on developing an energy project. The simulation model carried out a causal loop diagram (CLD) as the dynamic behaviour hypothesis, presented in Fig. 4.4, to develop a stock and flows diagram. The SD model is adapted from the rework cycle (Cooper, 1980). The structure of the adapted rework cycle incorporates the interaction of tasks required to develop the project such as work to be done, work done correctly, work done, and undiscovered work. However, this task interaction is influenced by workload, schedule pressure, productivity, and work rate.

4.3.2 Model Validation

The SD model applied tests of consistency and dimensionality suggested by Barlas (1996) and Qudrat-Ullah and Seong (2010). The simulation model is validated according to the actual data of the energy project, as presented in Fig. 4.5. Thus, this figure uses the rework SD model for PM to simulate project tasks (actual data) and work actually done (simulation data).

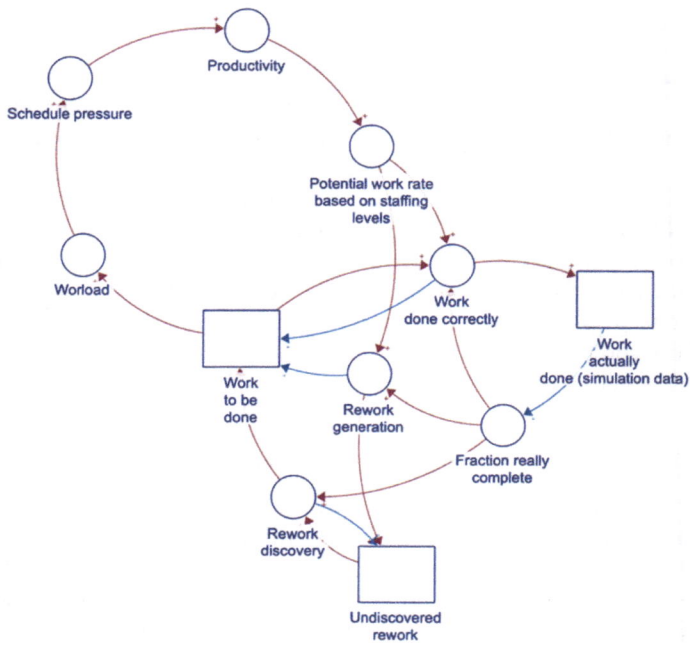

Fig. 4.4 Dynamic behaviour hypothesis

4.4   Results

The simulation results show the impact in sustainability dimensions in accordance to the triple bottom line (TBL) (Elkington, 1999), and administrative aspects (APM, 2019). The SD simulation model uses scenarios to contrast the influence of process innovation. That is to say, the sustainability dimensions are forecasted to determine the impact that an energy project will have over time, which process innovation acts as triggers to reduce project span.

4.4.1   Scenarios Design

The SD model used data from an energy project developed by the Colombian government's Ministry of Mines and Energy—FAZNI-GGC-521-IPSE-074-2017. The project took 18 months, from 10 November

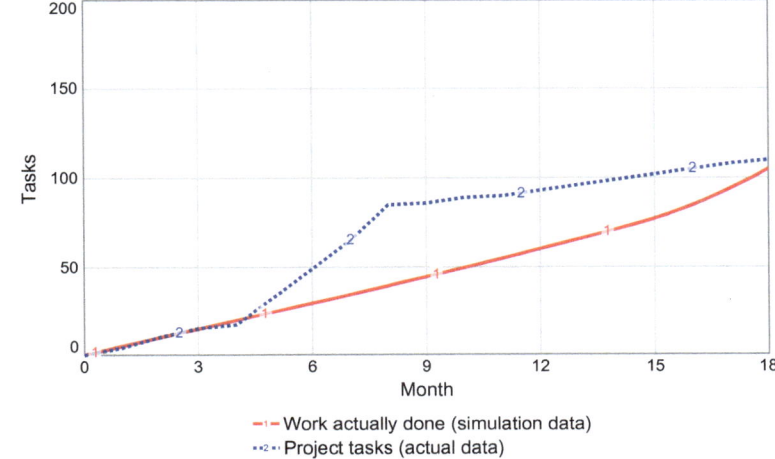

Fig. 4.5 Model validation test

2017 to 30 April 2019, and had a budget of USD 540.839. The SD model has been simulated within three scenarios which represent the process innovation to reduce project completion.

The first scenario represents actual project data—business as usual (BAU). The energy project was developed in 18 months including two delays due to public order which access to this region was restricted. The first delay started on 9 November 2018 and ended on 13 December 2018. The second delay started on 17 January 2019 and ended on 14 March 2018. The BAU scenario includes 3 months' time as a consequence on two delays.

The second scenario reduces three months to project termination produced by the two delays in the project. Reducing three months to project span represent 20% in process innovation. This scenario shows a simulation of project completion in 15 months.

The third scenario includes time reduction produced by paperwork and project delays. Actual project data shows that project paperwork was developed in 7 months which involves agreement, licence, contract, and environmental licencing according to Colombian legislation. According to authors analysis, it is possible to reduce this project paperwork from

7 to 4 months and avoid the 3 months project delay. To put it another way, the third scenario simulates the reduction in 6 months project span by increasing 50% process innovation.

4.4.2   Social Sustainability Dimension

Social aspects in this research focus on to provide energy access to a marginalized group. Developed countries involve technological capabilities, then, economic growth empathizes to produce goods and services for wealthy clients. By contrast, developing countries seem more plausible (Chataway et al., 2014). With this in mind, governments of developing countries must include and prioritize marginalized groups in their regions to ensure energy access for its community—SDG7 (SDG Report, 2019), even if developed countries do not have the issue of electricity lack. Significantly, ensuring access to energy for all requires project management. However, its management can go over budget and over time.

This SD model shows the simulation for 82 families as marginalized group in Colombia. The Colombian government through the project FAZNI-GGC-521-IPSE-074-2017 satisfies energy access to the community. The energy project offers an installed capacity of 62 kW for a constant consumption of 82 households PV adopters. However, Fig. 4.6 shows three scenarios in which project delivery could decrease project time span due to innovation process.

Figure 4.6 displays three scenarios.

The first scenario simulates the behaviour of the actual project, named as business as usual (BAU). The project took 18 months to be completed within project delivery. This scenario includes 25 years (300 months) in the back-end of the project due to the solar panel duty cycle (Jinko, 2015). Note the solar manufacturer warranties at maximum of 97.5% power performance in the first month and then, the power performance has a linear decrease at 83.1% power performance until 25 years linear power.

The second scenario is based on what if the project had not had a delay due to public order. The process innovation was estimated incorporating 20% to reduce project time span from 18 to 15 months. Meanwhile, the third scenario increases 50% in process innovation. This increase reduces from 18 to 12 months project time span. As a consequence, the final of back-end of the project will be in the month 312 for the third scenario.

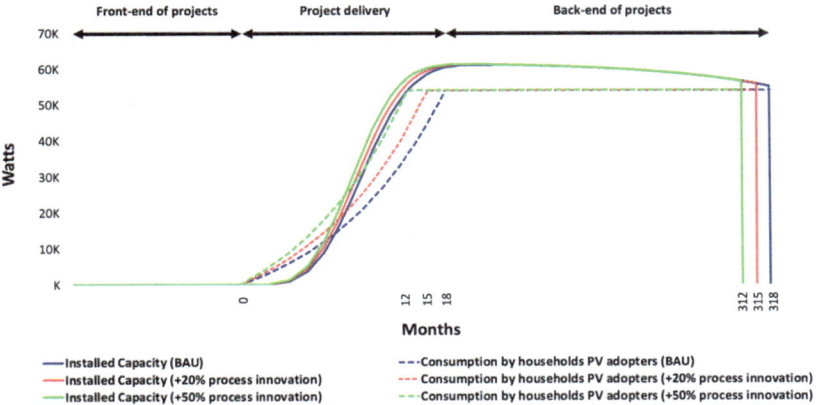

Fig. 4.6 Social sustainability results from a solar energy project

The three scenarios satisfy the social sustainability dimension providing energy access for inclusive community. Due to the solar energy project being sponsored by the Colombian government, this research does not include the origins as policies and government decision-making to order this project to the ministry. In addition, the provided data did not include front-end information of the project.

4.4.3 Environmental Sustainability Dimension

In light of the evidence that solar energy has low carbon dioxide (CO_2) emissions, three scenarios were developed for solar panels waste simulation after the back-end of the project, as shown in Fig. 4.7. After the 25 years at the end of solar panels useful life, the energy project will leave 186 solar panels out of service. To that end, the total height will be 363.826 metres (m). Put differently, that's the equivalent height of an Eiffel tower in solar modules.

By increasing process innovation to the solar energy project, three scenarios are shown in Fig. 4.7. If the project did not increase its installed capacity, energy access will be ended in month 318 for scenario 1, month 315 for scenario 2, and month 312 for scenario 3. For instance, to reduce negative impact on the environment, the project has to plan the management of solar panel disposal from the front-end of projects.

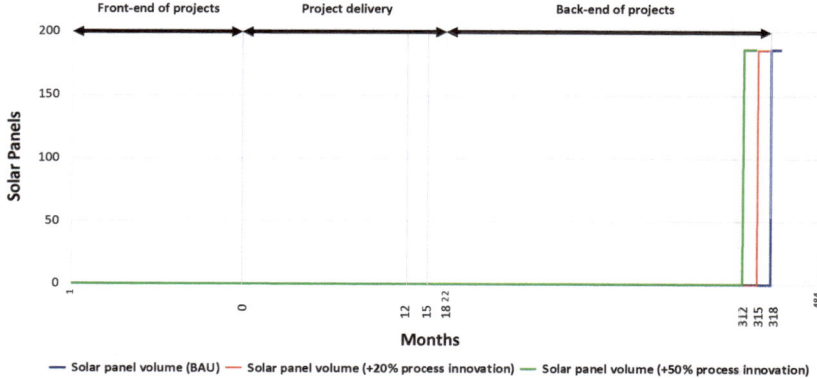

Fig. 4.7   Environmental sustainability results from a solar energy project

4.4.4   Economic Sustainability Dimension

The economic sustainability dimension has been always a concern for project managers to be on time and on budget which refers to project management success. What is not taken in consideration is project success related to the effectiveness and strategy-seeking benefits of the project (Cooke-Davies, 2007; de Wit, 1988). This is the case of providing access of energy using solar energy instead of fossil fuels. Even if the solar energy project has a fixed cost of USD 540.839, the project time span can be reduced by process innovation.

For this particular solar energy project case study developed by the Colombian government, a fixed cost is established in contract, other sponsorships can reduce menwork rate. However, the contract established that the project will be on time and on budget. The tariff solar photovoltaic (PV) scheme is also fixed by the Colombian government once the project is completed for maintenance, as shown in Fig. 4.8. The contract also establishes the monthly tariff will start in 18 months.

Figure 4.8 shows three scenarios for the economic sustainability dimension. The first scenario is BAU which the sum of tariff PV scheme, at the end of 25 years, will be USD 156 k paid by the final user. The second scenario is incorporating 20% process innovation which at the end the final user will pay USD 155 k at month 315. Meanwhile, the third scenario simulates an increase of 50% of process innovation with a total amount of USD 153 k at month 312. Even if the Colombian government

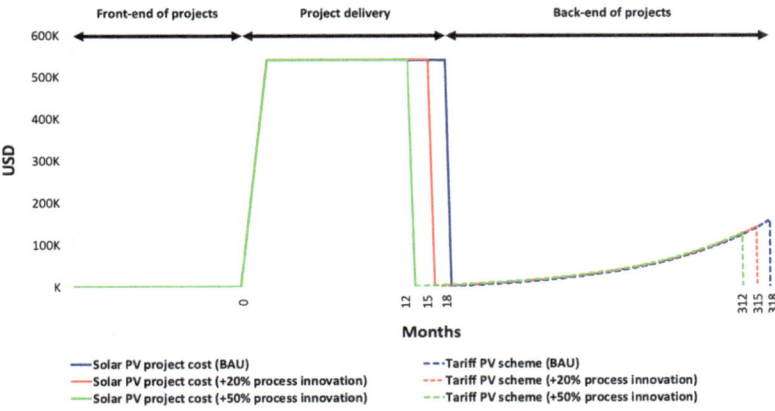

Fig. 4.8 Economic sustainability results from a solar energy project

has a fixed cost to develop the solar energy project, economic benefit will be for project's final user or families in the tariff PV scheme at the end of project life cycle.

4.4.5 Administrative with Ethics Sustainability Dimension

The administrative paperwork for the development of this solar energy project took 7 months distributed in 3 stages. First stage had two months' time for activities prior to contracting the work. Second stage had one month's time for contracting activities. The third stage took four months' duration for activities of supply and construction of works, and liquidation. The third scenario focuses on reducing administrative paperwork tasks by process innovation. This scenario was developed if paperwork manages its processes from 7 to 4 months. The administrative processes are different according to the sector, public or private. In order to have transparent processes, public sector seeks to warranty project competition by analysing companies' reputation. However, this analysis can increase time and paperwork, is the tool that public sector has.

The ethics for public and private sector are more related to the balance between profit and negative environmental impact. The results for this SD model show a clear procedure in obtaining the project data. The Colombian government's Ministry of Mines and Energy assisted with data recollection in the development of this simulation—Rad. 2019084512. All data was retrieved and validated according to the paperwork and project delivery.

4.5   DISCUSSION

Nowadays, PM requires the use of innovation to warrantee sustainability over time. However, sustainability expands project management success to project success. This research focuses in the application of how process innovation reduces project time span and has sustainability aspects and implications over time. The discussion of this research is based on a three-dimensional (3D) representation of the behaviour of projects due to innovation and sustainability illustrated in Fig. 4.9. The axes are time, project capability, and innovation level, as a result, a sustainability evaluation. The results section demonstrates the behaviour of time is linear; while innovation, sustainability, and PM are nonlinear and complex.

Project capabilities comprise tacit and explicit knowledge embodied in standardizing procedures, processes, tools, and guidebooks by learning level interaction. In addition, there is a close relation between project capabilities with innovation by an exploratory vanguard project, and an organization's resources (Brady & Davies, 2004). However, capability refers to the project performance; as Repenning and Sterman (2001) explain, "The actual performance of any process depends on two factors: the amount of Time Spent Working and the Capability of the process used to do that work" (Repenning & Sterman, 2001; Sterman, 2000).

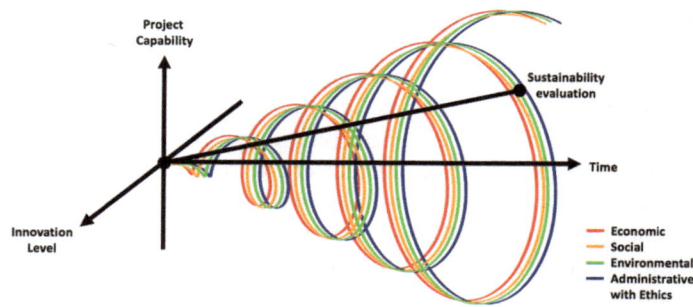

Fig. 4.9   3D representation of timeline proposal for innovation, sustainability, and PM

The results in developing tasks measure the project capability dimension. Besides this, once the project is completed, it will have the capacity to produce (e.g. numbers of products, amount of energy) (Herrera et al., 2019), or to use (e.g. infrastructure, passenger transport).

The purpose of the innovation level is to measure a project's innovativeness by equipment and materials, methods and techniques, and the application of new ideas during the planning and managing of the project (Lewis-Beck, 1977; Park et al., 2004). The innovation level dimension is the degree of innovativeness within the project (i.e. adaptation-innovation inventory) (Kirton, 1976). Thus although there are uncertainties and risks associated with the processes of innovation and PM there are opportunities for new capabilities (Davies et al., 2018).

SD started to develop economic-related models for industrial systems, social-related ones for social sciences, and environmental-related ones for controlled experiments (Forrester, 1961). These models are related to the three pillars of sustainability, to simulate complex and nonlinear behaviour (Nabavi et al., 2016). The differences in initial conditions and how these conditions change over time will contribute to a better assessment of the sustainability. These differences are a rate of change, positive or negative, influenced by the interaction between systems over time.

Besides project capability and innovation level are essential for project management success, its sustainability implications must be analysed in long-term future for project success. This research is based on four dimensions for sustainability in PM: social, environmental, economic, and administrative with ethics. Social dimension in this research relates to satisfy the seventh sustainable development goal (SDG), ensuring access to affordable, reliable, sustainable, and modern energy (SDG Report, 2019) by the development of the solar energy project as a case study. That is to say, the management of this project includes contributions to sustainable communities (Taylor, 2011), financial, construction social action, relationships with the local community (Martens & Carvalho, 2014), and societal interests (Silvius & de Graaf, 2019).

The management of projects extends its boundaries to understand the implication that a project will have in the long term. For instance, balancing public sector and society needs must be taken into consideration. On one hand, what projects governments of developing countries want to be developed. On the other hand, what inclusive community issues need to be solved. These two arguments need to be interlinkage with a close relationship between government and society, even more, for an inclusive community.

Planning in long term assures that energy issue will be finished in the future. To solve this issue, project delivery must consider population increase from the front-end. To put it another way, the project must satisfy installed capacity to the consumption by households over a long-term period. Before project start, in this case is required a relationship between government and society. This relationship can be managed by governance of innovation and inclusive innovation (Sandra Schillo & Robinson, 2017), however, it is used for policymaking instead of PM.

Good practices of the environmental sustainability dimension help to mitigate climate change (Morris, 2017). Solar energy could be tucked in consideration to contribute for the environment and be aligned with SDG13—climate change action. Developing countries are adopting renewable energies as strategy to satisfy society needs and be environmentally friendly, even if usually 25 years are its duty cycle. What is not taken in consideration is the number of solar panels at the end of its use. Thinking from the front-end of projects, it is possible to plan another use of this number of solar panels into another project. For instance, using this amount of solar panels for raw materials for infrastructure projects helps reuse and recycling of materials contamination and pollutants, disposal and processing (Taylor, 2011), impacts on the environment (Martens & Carvalho, 2014), and SDG9—build resilient infrastructure, promote sustainable industrialization, and foster innovation. Coupled with planning environmental decisions within front-end of projects, economic sustainability dimension in the long term can be positive.

One of the constraints in PM is costs. To warranty economic sustainability in project management success, final users, in this case, inclusive community must be included. Public sector must analyse project success which the long-term planning must study monthly payment by users. For instance, fossil fuels are attractive for government projects due to initial costs in comparison to solar technology. However, the final user will pay more in maintenance and fossil fuel supplies such as petrol. In addition, economic dimension for project success has financial benefits of good practices (Martens & Carvalho, 2014), SDG11—Make cities inclusive, safe, resilient, and sustainable, and SDG8—Promote inclusive and sustainable economic growth, employment and decent work for all.

Administrative with ethics sustainability dimension implies good practices and avoids corruption and personal interests at strategy, operational, and tactic levels, not to mention be aligned within SDG16—promote just, peaceful, and inclusive societies. Public and private sectors must have benefits for the society and the environment. It is not justifiable to make profits compromising or spending all of our natural resources. Its high price will have negative consequences in the future, for instance, climate change. In view of future negative consequences, administrative procedures must have ethical actions and decisions. Significantly, developing countries have higher negative consequences doing corrupted procedures at long-term range. To give an illustration of what this research found, projects can be compromised in the delivery side as delays and over costs. On the back-end of projects side, higher tariff scheme for final users/ community and negative environmental impact.

The lens of including innovation and sustainability to PM supports project strategy and concept. However, linking these disciplines requires complexity. To achieve project complexity, SD is used in this research. Sustainability can be managed for the project and after the project. Sustainability for the project is related to warranty project completion with. Furthermore, sustainability after the project seeks project's impact over time, for instance, consequences in social, environmental, economic, and administrative aspects. One key element demonstrated by using the SD simulation model is process innovation. For instance, innovation in processes allows to reduce project time span and mitigates uncertainties over time.

4.6 Conclusions and Future Work

A simulating model helps to public and private sector as a tool to understand sustainability behaviour caused by developing a project, in this case, a solar energy project. Having an SD model, variables can change, and results are given in real time. In addition, project managers and policymakers can use simulation models to forecast project's impact at long-term range for decision-making.

The SD model establishes process innovation as a key element for PM and sustainability. The SD model simulates scenarios by changing process innovation percentage. Increasing process innovation reduces project span; however, changing project deadline can produce delays due to fatigue in schedule pressure.

Simulating solar projects reinforces advantages in renewable energy. For instance, positive impact on social and environmental aspects. Even though economic aspects have initial costs, solar projects have benefit at long term. Furthermore, ethics are added to administrative sustainability dimension to avoid economic and personal interest. The administrative aspects include agreement, licence, contract, and environmental licencing which innovation in these processes will reduce delays, claims, and project span.

The research can be improved by simulating other type of projects within social, environmental, economic, and administrative sustainability aspects within the simulation model to understand system behaviour caused by the project. Social sustainability aspects can include jobs creation, health in the community, and contributions to sustainable communities (Martens & Carvalho, 2014; Taylor, 2011). Environmental sustainability aspects can include materials selection, climate change adaption, and waste avoidance, disposal, and processing (Martens & Carvalho, 2014; Taylor, 2011). Economic sustainability aspects can include Financial benefits of good practices, procurement strategies, and taxes and fines (Martens & Carvalho, 2014; Taylor, 2011). Administrative with ethics sustainability aspects can include legal procedures, culture beliefs and values, and public corruption laws.

Acknowledgements This work was supported by the Colombian Ministry of Science, Technology, and Innovation—Minciencias Scholarship Program No. 756, and by the Colombian Army Resolution No. 330. The authors would like to acknowledge receipt of funding from the Universidad Militar Nueva Granada (Grant, IMP-ECO-3402).

References

Alasad, R., & Motawa, I. (2015). Dynamic demand risk assessment for toll road projects. *Construction Management and Economics, 33,* 799–817. https://doi.org/10.1080/01446193.2016.1143561

Ansari, R. (2019). Dynamic simulation model for project change-management policies: Engineering project case. *Journal of Construction Engineering and Management, 145,* 22. https://doi.org/10.1061/(asce)co.1943-7862.000 1664

APM. (2019). *APM body of knowledge* (7th ed.). Association for Project Management, Buckinghamshire.

Barlas, Y. (1996). Formal aspects of model validity and validation in system dynamics. *System Dynamics Review, 12*, 183–210. https://doi.org/10.1002/(SICI)1099-1727(199623)12:3%3c183::AID-SDR103%3e3.0.CO;2-4

Bjorvatn, T., & Wald, A. (2018). Project complexity and team-level absorptive capacity as drivers of project management performance. *International Journal of Project Management, 36*, 876–888. https://doi.org/10.1016/j.ijproman.2018.05.003

Brady, T., & Davies, A. (2004). Building project capabilities: From exploratory to exploitative learning. *Organization Studies, 25*, 1601–1621. https://doi.org/10.1177/0170840604048002

Carvalho, M. M., & Rabechini, R. (2017). Can project sustainability management impact project success? An empirical study applying a contingent approach. *International Journal of Project Management, 35*, 1120–1132. https://doi.org/10.1016/j.ijproman.2017.02.018

Chataway, J., Hanlin, R., & Kaplinsky, R. (2014). Inclusive innovation: An architecture for policy development. *Innova Dev, 4*, 33–54. https://doi.org/10.1080/2157930x.2013.876800

Chronéer, D., & Backlund, F. (2015). A holistic view on learning in project-based organizations. *Project Management Journal, 46*, 61–74. https://doi.org/10.1002/pmj.21503

Cooke-Davies, T. (2007). Project success. In P. W. G. Morris & J. K. Pinto (eds.), *The Wiley guide to project, program, and portfolio management*. John Wiley & Sons, Inc., Hoboken, N.J.

Cooper, K. G. (1980). Naval ship production – A claim settled and a framework built. *Interfaces (Providence), 10*, 20–36. https://doi.org/10.1287/inte.10.6.20

Davies, A. (2019). Project management for large, complex projects. *APM Collaborating for Results, 1*–23.

Davies, A., Manning, S., & Söderlund, J. (2018). When neighboring disciplines fail to learn from each other: The case of innovation and project management research. *Research Policy, 47*, 965–979. https://doi.org/10.1016/j.respol.2018.03.002

de Wit, A. (1988). Measurement of project success. *International Journal of Project Management, 6*, 164–170. https://doi.org/10.1016/0263-7863(88)90043-9

Elia, G., Margherita, A., & Secundo, G. (2020). Project management canvas: A systems thinking framework to address project complexity. *International Journal of Managing Projects in Business*. https://doi.org/10.1108/IJMPB-04-2020-0128

Elkington, J. (1999). *Cannibals with forks: The triple bottom line of 21st century business*. Capstone.

Fernandes, G., O' Sullivan, D., Pinto, E. B., et al. (2020). Value of project management in university–industry R&D collaborations. *International Journal of Managing Projects in Business, 13*, 819–843. https://doi.org/10. 1108/IJMPB-08-2019-0191

Forrester, J. W. (1961). Industrial dynamics. 197.

Herrera, M. M., Dyner, I., & Cosenz, F. (2019). Assessing the effect of transmission constraints on wind power expansion in northeast Brazil. *Utilities Policy, 59*, 100924. https://doi.org/10.1016/j.jup.2019.05.010

Jinko. (2015). Eagle 72P.

Kirton, M. (1976). Adaptors and innovators: A description and measure. *Journal of Applied Psychology, 61*, 622–629. https://doi.org/10.1037/0021-9010.61. 5.622

Lane, D. C., & Husemann, E. (2008). Steering without Circe: Attending to reinforcing loops in social systems. *System Dynamics Review, 24*, 37–61. https://doi.org/10.1002/sdr.396

Lenfle, S., Le Masson, P., & Weil, B. (2016). When project management meets design theory: Revisiting the Manhattan and Polaris projects to characterize 'Radical Innovation' and its managerial implications. *Creativity and Innovation Management, 25*, 378–395. https://doi.org/10.1111/caim. 12164

Lewis-Beck, M. S. (1977). Influence equality and organizational innovation in a Third World Nation: An additive-nonadditive model. *American Journal of Political Science, 21*, 1. https://doi.org/10.2307/2110444

Lippe, S., & Vom Brocke, J. (2016). Situational project management for collaborative research projects. *Project Management Journal, 47*, 76–96. https://doi.org/10.1002/pmj.21561

Lopes, J. D., Braga, J. L., & Resende, M. A. (2015). Systems dynamics model for decision support in risk assessment in software projects. *Journal of Software: Evolution and Process, 27*, 976–989. https://doi.org/10.1002/smr.1754

Maniak, R., & Midler, C. (2014). Multiproject lineage management: Bridging project management and design-based innovation strategy. *International Journal of Project Management, 32*, 1146–1156. https://doi.org/10.1016/j. ijproman.2014.03.006

Martens, M. L., & Carvalho, M. M. (2014). A conceptual framework of sustainability in project management. *Project Management Institute*, 1–10.

Martens, M. L., & Carvalho, M. M. (2016). Sustainability and success variables in the project management context: An expert panel. *Project Management Journal, 47*, 24–43. https://doi.org/10.1177/875697281604700603

Martens, M. L., & Carvalho, M. M. (2017). Key factors of sustainability in project management context: A survey exploring the project managers' perspective. *International Journal of Project Management, 35*, 1084–1102. https://doi.org/10.1016/j.ijproman.2016.04.004

Matinheikki, J., Artto, K., Peltokorpi, A., & Rajala, R. (2016). Managing inter-organizational networks for value creation in the front-end of projects. *International Journal of Project Management, 34*, 1226–1241. https://doi.org/10.1016/j.ijproman.2016.06.003

Mishra, A., & Browning, T. R. (2020). Editorial: The innovation and project management department in the Journal of Operations Management. *Journal of Operations Management, 66*, 616–621. https://doi.org/10.1002/joom.1111

Morris, P. W. G. (2017). *Climate change and what the project management profession should be doing about it – A UK Perspective.* Association for Project Management, Princes Risborough, England.

Müller, R., Glückler, J., & Aubry, M. (2013). A relational typology of project management offices. *Project Management Journal, 44*, 59–76. https://doi.org/10.1002/pmj.21321

Nabavi, E., Daniell, K. A., & Najafi, H. (2016). Boundary matters: The potential of system dynamics to support sustainability? *Journal of Cleaner Production.* https://doi.org/10.1016/j.jclepro.2016.03.032

Nasirzadeh, F., Afshar, A., Khanzadi, M., & Howick, S. (2008). Integrating system dynamics and fuzzy logic modelling for construction risk management. *Construction Management and Economics, 26*, 1197–1212. https://doi.org/10.1080/01446190802459924

Pargar, F., Kujala, J., Aaltonen, K., & Ruutu, S. (2019). Value creation dynamics in a project alliance. *International Journal of Project Management, 37*, 716–730. https://doi.org/10.1016/j.ijproman.2018.12.006

Park, M., Nepal, M. P., & Dulaimi, M. F. (2004). Dynamic modeling for construction innovation. *Journal of Management in Engineering, 20*, 170–177. https://doi.org/10.1061/(asce)0742-597x(2004)20:4(170)

Pich, M. T., Loch, C. H., & De Meyer, A. (2002). On uncertainty, ambiguity, and complexity in project management. *Management Science, 48*, 1008–1023. https://doi.org/10.1287/mnsc.48.8.1008.163

Poziomek, E. J., Rice, D. W., Andersen, D. F. (1977). Management by objectives in the R&D environment – A simulation. *IEEE Transactions on Engineering Management, 24*, 45–51. https://doi.org/10.1109/TEM.1977.6447334

Qudrat-Ullah, H., & Seong, B. S. (2010). How to do structural validity of a system dynamics type simulation model: The case of an energy policy model. *Energy Policy, 38*, 2216–2224. https://doi.org/10.1016/j.enpol.2009.12.009

Repenning, N. P., Sterman, J. D. (2001). Nobody ever gets credit for fixing problems that never happened: Creating and sustaining process improvement. *California Management Review, 43*, 64–+. https://doi.org/10.2307/41166101

Roberts, E. B. (1964). *The dynamics of research and development.* Harper & Row.

Sadabadi, A. T., Kama, N. (2014). A conceptual-operative framework for in-process decision support of software project management practice. *International Journal of Software Engineering & Applications, 8*, 287–302. https://doi.org/10.14257/ijseia.2014.8.1.25

Sandra Schillo, R. M., & Robinson, R. (2017). Inclusive innovation in developed countries: The who, what, why, and how. *Technology Innovation Management Review, 7*, 34–46. https://doi.org/10.22215/timreview1089

SDG Report. (2019). SDG 7 – Affordable and clean energy.

Shen, L. Y., Wu, Y. Z., Chan, E. H. W., & Hao, J. L. (2005). Application of system dynamics for assessment of sustainable performance of construction projects. *Journal of Zhejiang University Science A, 6*, 339–349. https://doi.org/10.1631/jzus.2005.A0339

Silvius, A. J. G., & de Graaf, M. (2019). Exploring the project manager's intention to address sustainability in the project board. *Journal of Cleaner Production, 208*, 1226–1240. https://doi.org/10.1016/j.jclepro.2018.10.115

Sterman, J. D. (2000). *Business dynamics: Systems thinking and modeling for a complex world*. McGraw-Hill.

Taylor, T. (2011). *Sustainability interventions – For managers of projects and programmes*. Dashdot Publications.

United Nations. (2020). *The sustainable development goals report 2020*.

UPME. (2021). Unidad de Planeación Minero Energética UPME.

van Eck, N. J., & Waltman, L. (2010). Software survey: VOSviewer, a computer program for bibliometric mapping. *Scientometrics, 84*, 523–538. https://doi.org/10.1007/s11192-009-0146-3

van Oorschot, K. E., Eling, K., & Langerak, F. (2018). Measuring the knowns to manage the unknown: How to choose the gate timing strategy in NPD projects. *Journal of Product Innovation Management, 35*, 164–183. https://doi.org/10.1111/jpim.12383

The Sustainable Transition of the Public Bus Fleet in the City of São Paulo Until 2038

Tainara Volan, Mauricio Uriona Maldonado[ID]*,*
and Caroline Rodrigues Vas

Abstract In 2018, the city of São Paulo, Brazil enacted a law setting ambitious goals for the city's public bus transport CO_2, particulate matter, and NO_x with a deadline for reaching the target in January 2038. This encompasses a rapid technological transition of diesel-based buses towards cleaner options such as zero-emission or low emission vehicles. However, for this transition to take place, decision-makers need to be aware of the technologies combination that can bring them closer to the

T. Volan
Graduate Program in Production Engineering (PPGEP), Federal University of Santa Catarina (UFSC), Florianopolis, SC, Brazil

M. Uriona Maldonado (✉) · C. Rodrigues Vas
Department of Industrial and Systems Engineering, Federal University of Santa Catarina (UFSC), Florianopolis, SC, Brazil
e-mail: m.uriona@ufsc.br

C. Rodrigues Vas
e-mail: caroline.vaz@ufsc.br

M. M. Herrera (ed.), *Business Model Innovation for Energy Transition*, Palgrave Studies in Democracy, Innovation, and Entrepreneurship for Growth, https://doi.org/10.1007/978-3-031-34793-1_5

legislation imposed. Therefore, this paper seeks to assess the goal imposed by the law, analysing the interaction between biodiesel, electric, hybrid, biomethane, and natural gas technologies to achieve reductions in CO_2, PM, and NO_x.

Keywords Public Transport · Greenhouse Gases · Clean Energy Vehicles · System Dynamics · Decarbonization · Sustainability Transitions

5.1   INTRODUCTION

Reducing greenhouse gas emissions and mitigating climate change, as well as improving the well-being of the population, are essential items for a sustainable future. These actions directly influence the search for the achievement of sustainable development objectives, which are characterized as a universal call for action against poverty, protection of the planet, and guaranteeing peace and prosperity to society (ONU, 2019). According to the National Energy Balance (NEB, 2018), the transport sector is the second sector that most demands energy in Brazil and the main source of carbon dioxide emissions (CO_2) (EPE, 2018). Also, it is responsible for the emission of various pollutants harmful to health and degrading the urban environment (De carvalho, 2011). Therefore, cities are increasingly evaluating and implementing sound policies to encourage the adoption of zero-emission vehicles (Slowik et al., 2018).

The city of São Paulo (Brazil) recently passed a law to promote the progressive reduction of CO_2 and other greenhouse gases emitted by its bus fleet, through the gradual introduction of more sustainable and cleaner public transportation. It is a means to reach SDS Goal 9,[1] 11,[2] and 13[3] (ONU, 2019).

It is known that only with combustion engine technologies will it not be possible to achieve the goals imposed by the law, transitions to cleaner engine technologies and the use of non-fossil fuels will be

---

[1] ODS 9—Build resilient infrastructures, promote inclusive and sustainable industrialization, and foster innovation.

[2] ODS11—Making cities and human settlements inclusive, safe, resilient, and sustainable.

[3] ODS13—Take urgent measures to combat climate change and its impacts.

necessary. Public transport operators will need to develop long-term acquisition strategies to plan for these transitions and ensure compliance with emission reduction targets.

Given the importance of the public transport fleet for mobility in São Paulo, as well as its impact on motor vehicle pollution, investments in buses are a key long-term strategy to meet the city's environmental and sustainability goals. Changes in fuel and bus fleet technologies aim at the dual objective of improving the quality of service provided to system users and reducing emissions of pollutants harmful to air quality in the city (Dallmann, 2019). This must lead participants of the traditional mobility market to evaluate their strategy for future growth.

Alternative fuels can be used to reduce carbon emissions in the public transport domain, aiding in complying with the newer policies for greenhouse gas emissions and urban air pollution in the public transport sector, especially in the bus sector.

In this context, as the main objective of the present work, it is intended to evaluate the goal imposed by Less Pollutant Urban Transportation Fleet Act (Law No. 16,802), with different forms of transition to alternative fuels to diesel and, from there, verify the reductions achieved for carbon dioxide emissions (CO_2), particulate material (PM), and nitrogen oxides (NO_x).

We use System Dynamics (SD) which is a simulation modelling method used to analyse complex social systems and to aid in policy design. Forrester (1968) also describes how the science of feedback behaviour in social systems, in which the use of computer simulation tools aims to expose the nature of models in the process of interpreting and extending the concepts of non-linear systems with multiple feedback over time. Using this approach, it is possible to model the entire system for transportation planning through strategic and policy analysis and as a support tool for decision-making (Shepherd, 2014). SD uses a system of differential equations that are solved by numerical methods, and are composed of state variables (stocks), and rates (flows), generating—in the solution of the systems of equations—dynamic and non-linear behaviour of the system (Vaz & Maldonado, 2016).

SD has already been used in other works related to urban mobility, an example is the work of Fontoura et al. (2019) which sought to analyse the influence of Brazilian policies on the transport system, focusing on environmental, economic, and traffic variables. The case study was in the metropolitan region of São Paulo and the policy analysed was the

"Brazilian Urban Mobility Policy". Based on the results of the simulations, the implementation of the policy is not sufficient to obtain satisfactory and expected results from the policy.

Another work of relevance is that of Wen and Bai (2017), where SD was used to simulate the impact of different strategies on energy consumption and carbon emissions from urban traffic, in a case study applied in Beijing. The model includes subsystems of urban traffic, population, economy, energy consumption, and carbon emissions, bringing different strategic policies under discussion.

In addition, Sayyadi and Awasthi (2017) developed a simulation model to assess regulatory policies for sustainable transport planning. Policies that involve consumers are also included in the modelling, such as sharing or renting private cars, which consequently affect public transport.

The literature highlights the need to study the impacts of different actions on public policies applied to the transport sector. In the models, different elements are interrelated, such as economy, infrastructure, number of vehicles, fuel consumption, and safety, among others. In the development of the models, different combinations of these elements are used, to represent the complexity of the transport system.

Therefore, we understand the importance of applying SD to assist in understanding the public transport system and in making decisions for the case of the city of São Paulo.

5.2 Theoretical Framework

Large sectors such as energy supply, water supply, or transportation can be conceptualized as sociotechnical systems (Markard et al., 2012). Such systems consist of networks of actors (individuals, firms, and other organizations) and institutions (social and technical norms, regulations, standards of good practice), as well as artefacts and material knowledge (Geels, 2004; Weber, 2003).

In these sociotechnical systems, transitions can occur, which initially referred to large-scale transformations in society or important subsystems, during which the structure of the social system changes fundamentally (Rotmans et al., 2001). Later, the definition shifted to a fundamental change in structure (organizations, institutions), culture (norms, behaviour), and practices (routines, skills) (Loorbach & Rotmans, 2010), i.e., the dominant way in which a societal need (transportation, energy, agriculture) is satisfied, this may take one or two generations (25–50 years) to achieve (Kemp & Loorbach, 2003).

These significant changes in society can be called transitions to sustainability, which are long-term, multidimensional, and fundamental transformation processes through which established sociotechnical systems shift to more sustainable modes of production and consumption (Markard et al., 2012). Such changes are governed by involving large social systems, not only focusing on technology but also on organizational structure and culture (Tukker & Butter, 2007). It is also considered a multifaceted social challenge, involving technological, political, and behavioural changes, precisely at the intersection of production, energy use, and transportation systems (Nilsson & Nykvist, 2016).

One example of sustainability transition is the ongoing transition of the electricity sector towards increasing the share of renewable energy, such as solar photovoltaics and wind power (Markard et al., 2020).

ICEs are part of large sociotechnical systems. Thus, the required change goes far beyond changing the engines and transmissions of cars but implies changes in the way cars are built and automotive value chains are organized. In addition, the electrification of transportation creates new infrastructures and interfaces between transportation and energy systems, as well as favouring new mobility concepts, consumer attitudes towards mobility, and vehicle ownership. Anticipating these systemic changes, new alliances are emerging between energy companies, automobile manufacturers, and software companies (Altenburg et al., 2015).

Sustainability transitions have three main characteristics that differ from other transitions, these are: orientation towards previously defined goals; innovations do not offer obvious benefits to users, implying policy, subsidy, and regulatory needs due to market failures; and they deal with lock-in mechanisms that create a path dependency that hinders incumbent firms' moves to new technologies (Geels, 2011).

In the mobility sector, the transition to EVs has clear objectives, which can be defined as the pursuit of the UN SDGs; if the transition occurs it may affect different objectives. As for the second characteristic, the need for policies, subsidies, and regulations, EVs are costly compared to ICEs of the same category. In this way, subsidies are one of the ways to increase the number of sales. Subsequently, the need for infrastructure change is presented, for example, investments in recharging stations, which consequently generate new regulations in the energy market. Finally, there is the characteristic related to major actors of the system, an example is the oil market, which directly influences the development of this innovation, because over time large investments were made, and still are, for oil exploration, a sociotechnical system already built and in operation becomes an inhibitor of innovations (Geels, 2011).

5.3   Public Transport in the City of São Paulo

This section includes a view of the public transportation system in the city of São Paulo with special focus on the Less Pollutant Urban Transportation Fleet Act (Law No. 16,802, of January 17, 2018).

The expansion of cities meant that public transport assumed an essential role in displacing people within the urban environment. Urban buses are the most widespread mode of public transportation worldwide. This is related to its flexibility, its ability to adapt to different demands, simple technology, ease of switching routes, and low cost of manufacturing, implementation, and operation when compared to other modes (Segantin, 2019).

São Paulo's public transport system operates with approximately 14,000 buses on 1340 lines. It is the largest system in Brazil, and it is among the largest bus fleets in the world (SPTrans, 2019). The fleet is mainly comprised of diesel fuel (98%), which does not have the best technologies available to control diesel engine emissions (Miller & Façanha, 2016) and they cover an area of 91,250 km annually on average (Raymundo & Reis, 2015).

In January 2018, the City Council of São Paulo established Less Pollutant Urban Transportation Fleet Act, which determines that transport service operators (Urban Passenger Transport) should promote the progressive reduction of emissions of carbon dioxide from fossil origin, and toxic pollutants emitted in the operation of their fleets, through the gradual use of cleaner and more sustainable fuels and technologies. They also state that the process of replacing vehicles and cleaner technologies will occur gradually and will naturally occur when replacing the batch of older vehicles that are removed from the fleet, according to the maximum permitted age contractual rules of vehicles.

The Act requires reductions in tail-pipe CO_2 emissions but does not regulate CO_2 emissions from producer to consumer (end-of-pipe), associated with the production and distribution of raw materials and fuels,

Table 5.1 Pollutant reduction targets

Pollutant	At the end of 10 years (%) (January 2028)	At the end of 20 years (%) (January 2038)
Fossil CO_2	50	100
PM	90	95
NO_x	80	95

Source Less Pollutant Act of the City Council of São Paulo

nor does it consider climate pollutants other than CO_2, such as methane, nitrous oxide, and black carbon (Dallmann, 2019).

The targets for reducing fossil CO_2 exhaust emissions, PM, and NO_x adopted in the Act, are 50, 90, and 80%, respectively, within 10 years after the Act comes into force. Within 20 years, 100, 95, and 95% must be achieved respectively. The values are summarized in Table 5.1.

As shown in Fig. 5.1 and serving as a reference mode for the analyses to be carried out, the emissions of the São Paulo fleet in 2016 were estimated at 1.24 million tons of CO_2/year. Based on this estimate, annual CO_2 emissions in the fleet must be reduced by 0.62 million tons/year to meet the 50% reduction target to the reference fleet within 10 years, and in the next 20 years fully eliminated. For PM, in 2016, 144.7 tons were issued, the target for 10 years is the value to reach 14.5 tons and in 20 years, 7.2 tons. For NOx, 9130 tons were issued in 2016, the target for 10 years is to reach 1830 tons, and in 20 years 460 tons (Dallmann, 2019).

5.4 MATERIAL AND METHODS

The model depicts the ageing chain of current fossil fuel buses, as shown in Fig. 5.2. There is no inflow of new buses on this chain, as we assume all new buses will necessarily be zero or low emissions.

The stocks are named level 1 for buses from 0 to 5 years, level 2 for buses from 6 to 15 years, level 3 for buses from 16 to 25 years, and level 4 for buses over 25 years. Scrapping rates for levels 1 and 2 are 1% (Automotive-Business, 2015), and 75% and 85% for levels 3 and 4 respectively.

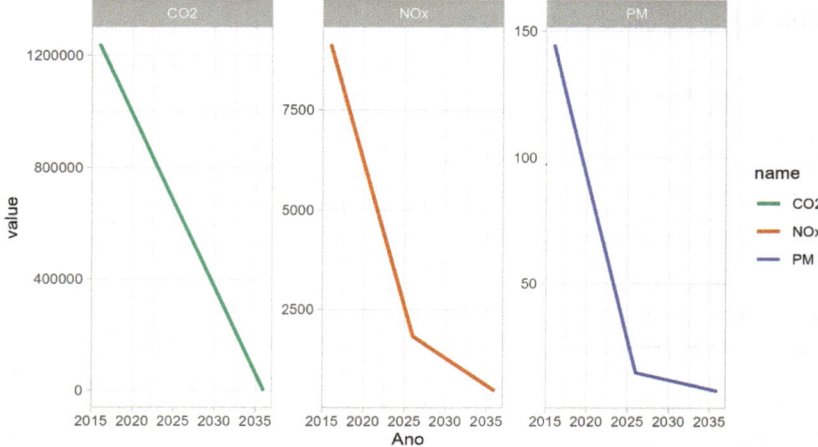

Fig. 5.1 Reference modes for emission of pollutants (*Source* Adapted from São Paulo Municipal Law 16.802)

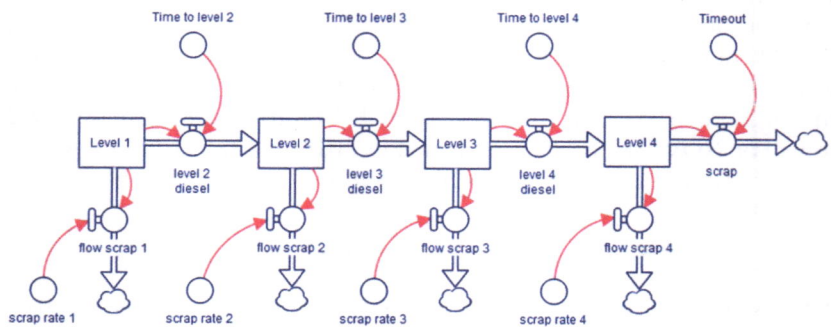

Fig. 5.2 Ageing chain of fossil fuel buses (*Source* The authors)

Next, we determine the equations for the exit of diesel buses for other chains, they follow the premise of Eq. 5.1.

$$Flow\,scrap\,1 = level\,1 \times scrap\,rate\,1 \qquad (5.1)$$

The flow from one stock to the next represents the ageing of fleet, and its modelled using Little's Law, in which in a system in equilibrium the average amount of time something takes to go through a process is equal to the number of units in the process divided by its average rate, thus following Eq. 5.2.

$$Level\ 2 = \frac{level\ 2\ diesel}{time\ to\ level\ 2} \tag{5.2}$$

There are several options of zero and low emission vehicles in the market, which offer various degrees of emissions improvement compared to the diesel buses. In this study, we focus on biodiesel, battery electrics, compressed natural gas (CNG), biomethane, and hybrid (electric and diesel).

As vehicles are being scrapped and leave the ageing chain in Fig. 5.2, they enter a new ageing chain of zero and low emission bus options. Figure 5.3 shows biodiesel and electric bus chains, a similar structure is used for the other alternatives, i.e., biomethane, CNG, and hybrid.

The variable "scrap flows" represents the sum of all outgoing stocks, both diesel and alternative fuel buses, later, this variable will be the system input. Together with it, there is the natural rate of increase of the transport system, which corresponds to a value close to 1% per year.

Each of the proposed alternatives emits a specific value of CO_2, PM, and NO_x. In Fig. 5.4 we show how emissions are calculated for biomethane buses. All the other vehicle emissions are calculated similarly.

The variable "total biomethane buses" is repeated for the other technologies and represents the total number of buses of that type in the model. The emission factor is then multiplied by the number of kilometres run in the year and the number of buses in the ageing chain, thus generating the gas stock, according to Eq. 5.3. This equation is replicated for the different fuels and different gases.

$$CO_2\ biomethane\ emission = biomethane\ emission\ factor \times average$$
$$km\ driven \times total\ biomethane\ buses$$

$$\tag{5.3}$$

Next, all emissions are summed up, having as the final value the total emissions for each pollutant (CO_2, PM, and NO_x). The parameters are summarized in Table 5.2.

The characteristics of the different types of buses are added into the model according to data from the literature. Thus, the CO_2 emission value for diesel buses is equivalent to an average of the different bus models, this value is approximately 1000 g/km (ANTP, 2016). The PM

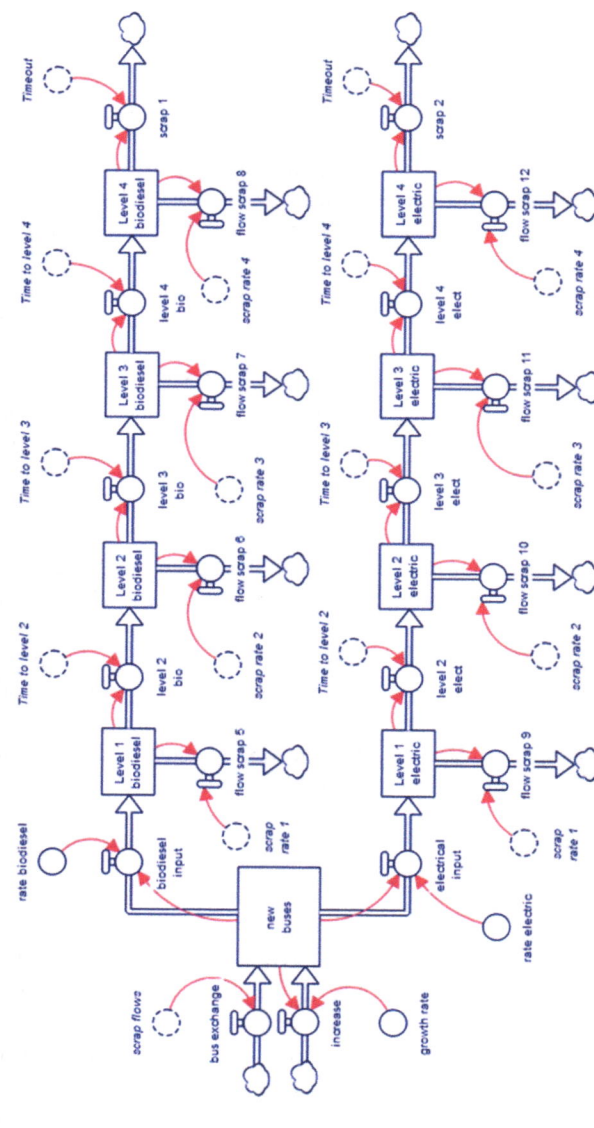

Fig. 5.3 Alternative bus ageing chain (*Source* The authors)

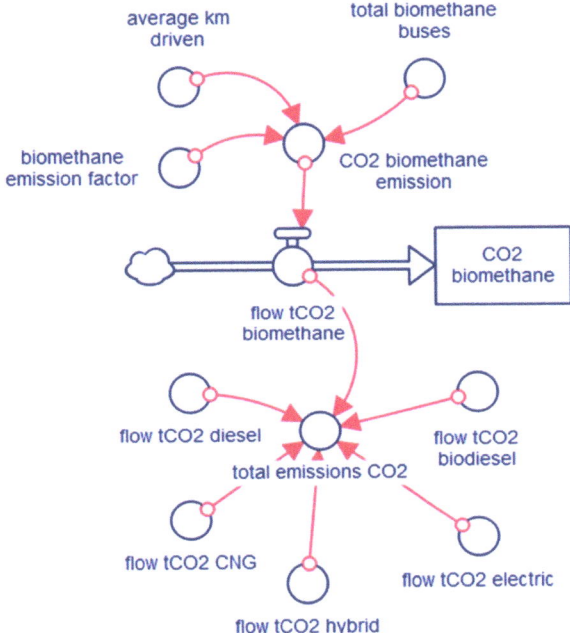

Fig. 5.4 CO$_2$ emissions estimation (*Source* The authors)

Table 5.2 Parameters used in the model

Parameters	Values
Average kilometres driven	91.250 km/year
Scrap rate—level 1 and level 2	1%
Scrap rate—level 3	75%
Scrap rate—level 4	85%
Growth rate	1%

Source The authors

emission is on average 0.115 g/km (ANTP, 2016). For NO$_x$ we also performed an average of the emission factor per g/km, of Euro III diesel, being a final value of 8 g/km (MMA, 2011).

Biodiesel can be used in internal combustion engines and replace, partially or totally, diesel coming from petroleum. It can reach 15.79%

less CO_2 emission than conventional diesel (only emissions in final use—fuel burning) (ANTP, 2016), which also brings a 26.8% reduction in PM emission. However, it increases the emission of NO_x by an average of about 8% (BIODIESELBR, 2011).

Biomethane can be produced from biogas generated by household, industrial, and agricultural waste. The use of biogas from waste for transportation has two major advantages: it replaces a fossil fuel and prevents the release of biomethane directly into the atmosphere, simply by treating and purifying the biogas to meet official specifications of 90–99% methane. This type of fuel demands a supply infrastructure and technology for its combustion (Falco, 2017). Biomethane emits on average 75% less PM than diesel (ABEGAS, 2019). It also emits 85% less CO_2 and 86% less NO_x (SIAMIG, 2018).

Hybrid buses combine a conventional internal combustion diesel engine with an electric propulsion system, so they are called hybrids. Electric propulsion aims to achieve greater fuel economy since electric motors are more efficient than internal combustion engines. They can be classified in parallel or series. In parallels, both the electric motor and the internal combustion engine are connected to the vehicle's mechanical transmission; in this technology, the electric motor usually transmits energy to move the wheels at low speeds, and the combustion engine enters when the vehicle acquires a higher operating speed (between 20 and 30 km/h). In series hybrids, only the electric motor drives the vehicle, so the combustion engine serves as a mini generator to recharge the batteries and drive the electric motor, which in turn drives the vehicle's wheels (Filho, 2011).

Hybrid buses have greater mechanical complexity than conventional diesel buses and vehicle acquisition and maintenance costs are higher; however, this can be compensated over the lifetime by great fuel economy (ANTP, 2016). They show a reduction of 80% of NO_x, 90% of PM, and 35% less CO_2 when compared to diesel (ANTP, 2016).

The CNG is one of the partially sustainable alternatives, due to its reduced local environmental impact, reduction of internal and external bus noise, availability, and competitive cost with diesel technology. Surrounded by advanced distribution and motorization technologies, CNG has stood out as an alternative in several countries, such as Madrid—Spain, Frankfurt—Germany, and Athens—Greece, among others (ANTP, 2016). They emit approximately 0.29 g/km of NO_x (MMA, 2013), and also reduce CO_2 emissions by approximately 37% (EPA, 2019) and 98% of PM emissions to diesel (SCANIA, 2018).

Electric buses reduce dependence on fossil fuels, eliminate greenhouse gas emissions because they do not emit exhaust gases such as fossil CO_2, PM, and NO_x, and also reduce noise in cities. Energy can be transferred to the bus in several ways: by charging the batteries at night in the garages, charging at the terminal (end or start of each line), and bus stops (Dallmann, 2019; Olsson et al., 2016). The autonomy can reach about 300 km for a medium load and 280 km for a full load, according to tests conducted in São Paulo. Electric buses already operate in the thousands in major cities in Asia, Europe, the United States, Japan, Colombia, and Mexico (ANTP, 2016).

Given the above, the quantities of gases emitted for each type of bus used in the simulation are summarized in Table 5.3.

After obtaining all the variables and parameters, we performed some tests to improve the model. The modelling process was documented, which allowed the replication and review of the results. In addition, the units of measurement for all variables and parameters were checked and the dimensional consistency of all equations was verified. Extreme condition tests were also performed to check whether the model behaves realistically. After performing these tests, we observed that the model shows realistic behaviour. Given this, we proceeded with the analysis of the scenarios in the next section.

Table 5.3 Gas emission (g/km)	Fuel	CO_2 (g/km)	PM (g/km)	NO_x (g/km)
	Diesel	1000	0.115	8
	Electric	0	0	0
	Biodiesel	842.10	0.08418	8.64
	Natural gas	630	0.0023	0.29
	Hybrid	650	0.0115	1.6
	Biomethane	150	0.02875	1.12

Source The authors

5.5   RESULTS AND DISCUSSIONS

Depending on the scrapping rates considered in the simulation, it is possible to observe the behaviour of the substitution of diesel buses. The transition to sustainable buses takes place slowly, and in 2040 it is still concentrated in the fleet of around 1500 buses, according to Fig. 5.5.

Thus, at the end of the period stipulated by law, there are still fossil fuel vehicles in the public fleet, which consequently causes emissions of CO_2 and other gases and cannot achieve zero emissions. This occurs because transitions are slow, the system must be reorganized to achieve the effectiveness of the service, and economic planning is necessary so that the costs do not reach the end customer. A way out of this bottleneck would be to speed up the process of fleet replacement so that in 20 years there is complete replacement.

For the next simulations, we use the above scenario as a basis. The scenarios were planned in such a way that there was a combination of most technologies available in the market, with two to three technologies dominating the market, being discarded the possibility of investing in a single alternative. This is due to the different characteristics of the technologies, which depend on the infrastructure itself. For buses that travel a longer mileage during the day, it is preferable to use fuels that give greater reach, this becomes a barrier for electric buses, which have limited reach and the need for their infrastructure for cargo that is not

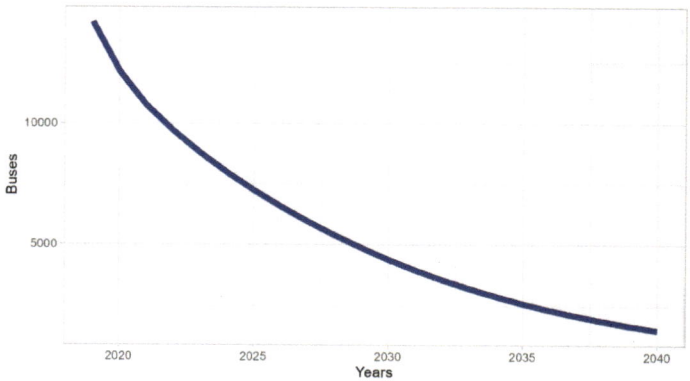

Fig. 5.5 Decrease of bus fleet to combustion (*Source* The authors)

Table 5.4 Scenarios for fleet replacement

Abbreviation	Combination	CNG (%)	Electric (%)	Hybrid (%)	Biodiesel (%)	Biomethane (%)
C1	Hybrid+Biodiesel	6	6	40	40	8
C2	CNG+Biomethane	40	6	6	8	40
C3	Electric+Hybrid	8	40	40	6	6
C4	Electric+Biomethane	6	40	8	6	40
C5	CNG+Hybrid+Biodiesel	30	5	30	30	5
C6	CNG+Electric	40	40	5	5	10
C7	Biomethane+Biodiesel	4	3	3	45	45
C8	Equal	20	20	20	20	20

Source The authors

common in Brazil, thus favouring the use for smaller routes, for example, and do small recharges at the terminals in the intervals between routes.

Table 5.4 shows the parameters for replacing the bus fleet for the different types of fuel.

The first scenario (C1) expresses the investment in hybrid buses and biodiesel. Hybrids are intermediates between electricity and combustion, a technology that does not depend on infrastructure, and biodiesel-powered buses are already a reality in many countries. In Brazil, the advantage comes from the large biodiversity of oilseeds for fuel production. Given this, 40% of investment in buses is simulated for each of them, for the other 6% in CNG and electricity, and 8% for biomethane.

The second scenario (C2) values CNG and biomethane as options. CNG has been available in Brazil since 1980, but consumption in the transportation sector is low, only 2% in 2016. As for biomethane, it can be leveraged by Renova-Bio, which establishes targets for reducing carbon intensity in the fuel matrix by 10% by 2028 and creates carbon credit trading mechanisms (Roitman & Da Silva, 2018). In addition, the infrastructure of these two fuels is similar and can be shared. Therefore, the simulation is based on 80% divided equally between CNG and biomethane, the rest being divided into other technologies.

For the third scenario, C3, the focus is on the shift to electric mobility, having a focus on electric and hybrid buses. According to the IEA (2018), more than 1 million electric cars were sold worldwide in 2017 and the world fleet already exceeds 3 million vehicles.

In Brazil, sales do not exceed 10,000 units, being an incipient technology. It is expected that sales will become higher when the cost of batteries is reduced. Consequently, given the technological evolution, and because these vehicles have better technology, performance, and energy efficiency, the demand for electrics will increase, leading to higher growth of the fleet (Roitman & Da Silva, 2018). Hybrids resemble electrics on a technological basis, so they may grow together in the market.

With 40% for electric bus and 40% for biomethane the fourth scenario (C4) was formed, the technologies were chosen because both presented the lowest values of CO_2 emission, the electric considered null (for exhaust) and the biomethane about 150 g/km, 85% lower than diesel, from the point of view of emission, would be the best environmentally.

Later, there is the fifth scenario, C5, which covers three technologies: CNG, hybrid, and biodiesel. These technologies, which have already been described above, together may become a dominant scenario, due to the existing knowledge about them and dominated by companies and researchers in Brazil.

In the case of the sixth scenario, C6, investments are in CNG and electricity. In this case, it is believed that as the CNG is already known by the actors of the sector, investments and greater acquisition of vehicles with this technology may occur, and combined with the electric one, which has significant benefits in the short and long term in the environmental issue, can together be driven.

The seventh scenario (C7) is biomethane and biodiesel. This combination meets the country's potential in producing these fuels. The last scenario, C8, has equal investment among the technologies, being elaborated to demonstrate a greater diversification of the fleet and identify the potential benefits for the user and the environment.

Returning to the targets imposed by Act, the reduction in CO_2 emissions should be 50% by 2028 (0.62 million tons) and in 2038 years should be zero. The results of the simulations for each scenario are presented in Fig. 5.6.

The goal of reducing CO_2 by 50% is achieved in 2029, by C4 (electric+biomethane), with approximately 602 thousand tons. However, in 2038 the value is not zero, because there are still diesel buses in the fleet and the biomethane emits polluting gases, however, the reduction is significant if compared to 2016 data, reaching a value close to 70%

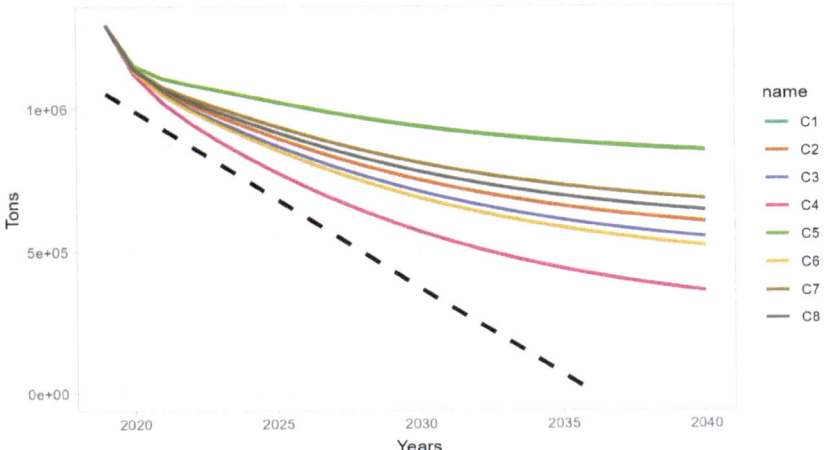

Fig. 5.6   Carbon dioxide emission—CO_2 (tons) (*Source* The authors)

reduction. Another scenario with significant results is C6, with a predominance of CNG and electric fuels, reaching 2028 approximately 743,673 tons of CO_2.

Scenarios C1 and C5, with a predominance of hybrid and biodiesel for the first and CNG, hybrid, and biodiesel for the second scenario, are the ones that present the worst results of CO_2 emission reduction. They are less aggressive technologies than diesel, however, still emit considerable CO_2 and do not meet the targets required by law. However, the C3 scenario, electric+hybrid, i.e., electric mobility, presents a median result, although electric buses predominate, hybrids occupy half of the fleet, which are more efficient than diesel, but still emit around 650 g/km of CO_2, higher emissions than CNG and biomethane.

For the PM, the reduction targets are 14.5 tons for 2028 and 7.2 tons for 2038. The results of the simulations for the different scenarios established are presented in Fig. 5.7.

The value of the PM target imposed by the law is not reached by any scenario, what comes closest is the C6, with a combination of CNG and electricity, presenting a reduction of approximately 53% for the first 10 years and in 20 years a reduction of 83%. These two technologies present the best emission rates among those analysed, thus, together they

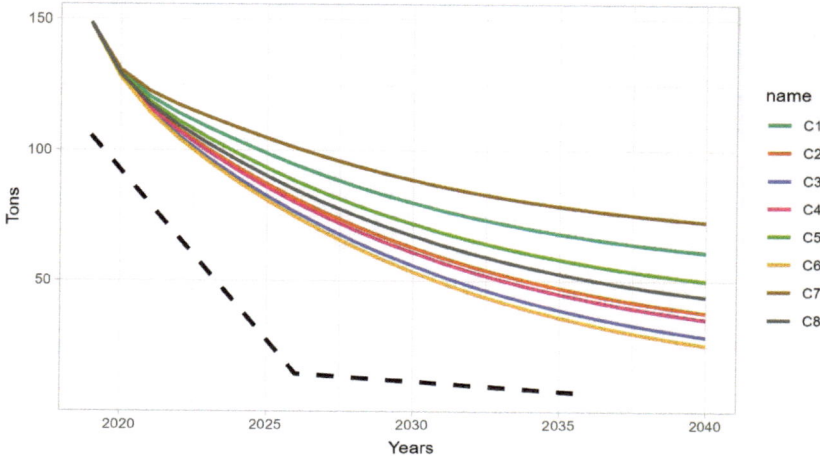

Fig. 5.7   Emission of Particulate Matter—PM (tons) (*Source* The authors)

have the best result. Next, there is the C3 scenario (electric and hybrid), with 55 and 75% PM reduction, for 2028 and 2038, respectively.

The scenario that presented the worst result is C7 (biodiesel + biomethane), which reduces on the year 2038 only 48% of the base value of 2016, this result comes from the PM emission factor, for the technologies analysed are the two largest emitters after diesel. The C1 scenario (hybrid and biodiesel) also presents a low performance, reducing in 20 years only 56% of emissions.

For analysis of the NOx results, we also consider the target imposed by law, which is 1830 tons for 2028 and 460 tons in 2038. The results based on the scenarios are presented in Fig. 5.8.

Again, faced with the simulations carried out, none of the scenarios presents the achievement of the goals imposed by law. The closest scenario is C6, a combination of CNG+electric, with reductions close to 50% in 2028 and 76% in 2038. Similar to the previous simulation, the worst results are the combination of biodiesel and biomethane (C7) and the C1 scenario, which consists of hybrid models and biodiesel, which in 2028 will reduce an average of 23% and for the year 2038 the value of 34%. The rates used in the study, which are the highest concerning diesel, are also due.

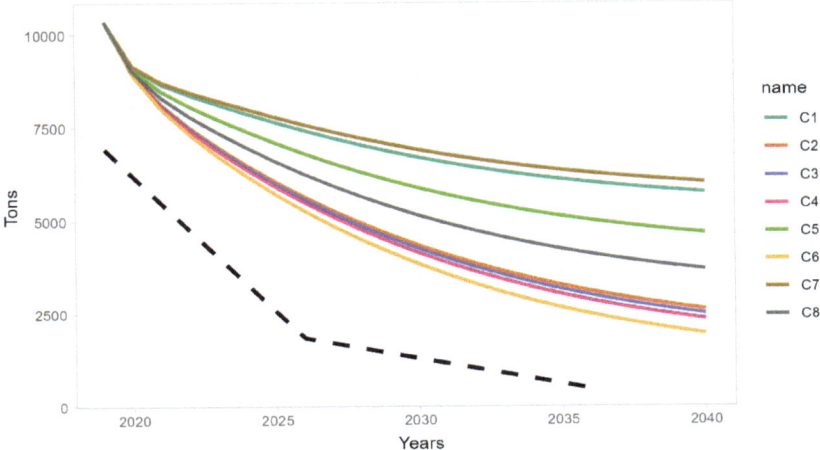

Fig. 5.8 Nitrogen oxide emission—NO_x (tons) (*Source* The authors)

Given the simulations carried out, the best combinations of alternatives for diesel-powered buses included the use of electric buses. The emissions for the different pollutants analysed were considered null, considering only the exhaust emission. In Brazil, the energy matrix is predominantly hydroelectric, which helps in the low pollution of electric buses, but it is necessary to take into consideration the thermoelectric plants or other sources that can be used, which can emit pollutants.

However, although in two cases the target will not be achieved, in any selected fleet combination, the substitution of diesel by other types of technologies, results in reductions in emissions in the range of 51% CO_2, 70% PM, and 64% NO_x by 2040, which corresponds to a local environmental benefit of extreme importance for public health, given the large total mileage travelled by public buses and their proximity to the population exposed to these emissions.

5.6   Conclusion

According to the sustainability transitions theory, transitions refer to large-scale transformations in society or important subsystems, during which the structure of the social system changes fundamentally. These changes are usually slow and require enormous efforts from different stakeholders in order to succeed.

The study showed one of the most radical sustainability laws in Brazil and the potential outcomes of switching fossil fuel buses of the public transportation system in the city of São Paulo for zero or low emission ones over a period of two decades.

The proposed scenarios were based on existing technologies and were organized to reflect possible choices that would be made by those responsible for operating the bus system, once favourable conditions were created for the introduction of zero or low emission technologies.

Given the results found in our simulations, the mix with best results are electric and biomethane, for CO_2 reduction, and CNG and electric buses for PM and NO_x reductions. Although, as shown in our results, none of the policy scenarios is able to fully comply with the Act, therefore suggesting that a stronger enforcement should be in place (such as enforcing faster substitution) if the City desires to achieve the goals imposed by the Act.

The emission factors of the polluting gases used for the simulation are preferably illustrative and not conclusive, as there may be data considering life cycle emissions and not just exhaust emissions. In addition, we stress that the analysis made in this work is simplified, considering only the benefits caused by the reduction of pollutant gas emissions. Thus, for future work, the insertion of cost variables is suggested, since it is necessary that the decision-makers responsible for bus changes also have the economic vision of each technology and thus make it efficient both economically and environmentally.

References

ABEGAS. (2019). *Ônibus movido a gás natural e biometano será testado em Curitiba* [Online]. Available: https://www.abegas.org.br/arquivos/70934. Accessed September 2019.

Altenburg, T., Schamp, E. W., & Chaudhary, A. (2015). The emergence of electromobility: Comparing technological pathways in France, Germany, China and India. *Science and Public Policy, Oxford University Press (OUP)*, 43(4), 464–475. https://doi.org/10.1093/scipol/scv054

ANTP. (2016). *Impactos ambientais da substituição dos ônibus urbanos por veículos menos poluentes*. In PÚBLICOS, A. N. D. T. (Ed.).

Automotive-Business. (2015). *Frota Circulante atinge 41,5 milhões de veículos* [Online]. Available: http://www.automotivebusiness.com.br/noticia/21922/frota-circulante-atinge-415-milhoes-de-veiculos. Accessed September 2019.

BIODIESELBR. (2011). *Emissão de poluentes atmosféricos locais do biodiesel em comparação com o diesel mineral* [Online]. Available: https://www.biodieselbr.com/efeito-estufa/gases/emissoes. Accessed September 2019.

Dallmann, T. (2019). Benefícios de tecnologias de ônibus em termos de emissões de poluentes do ar e do clima em São Paulo. In Transportation, I.-T. I. C. O. C. (Ed.), *ICCT—The International Council on Clean Transportation*. ICCT—The International Council on Clean Transportation.

De Carvalho, C. H. R. (2011). *Emissões relativas de poluentes do transporte motorizado de passageiros nos grandes centros urbanos brasileiros*. Texto para Discussão, Instituto de Pesquisa Econômica Aplicada (IPEA).

EPA. (2019). *Oil and gas extraction effluent guidelines* [Online]. United States Environmental Protection Agency. Available: https://www.epa.gov/eg/oil-and-gas-extraction-effluent-guidelines. Accessed October 2019.

EPE. (2018). *Balanço Energético Nacional 2018: Ano base 2017*. Empresa de Pesquisa Energética.

Falco, D. G. (2017). *Avaliação do desempenho ambiental do transporte coletivo urbano no estado de São Paulo: Uma abordagem de ciclo de vida do ônibus a diesel e elétrico à bateria*. Universidade Estadual de Campinas—UNICAMP.

Filho, A. F. M. (2011). Avaliação do ciclo de vida de diferentes tecnologias de ônibus: Eficiência energética e emissões de poluentes em operação real. *Rede C40 Cities (Grupo das Grandes Cidades líderes pelo Clima)*.

Fontoura, W. B., Chaves, G. D. L. D., & Ribeiro, G. M. (2019). The Brazilian urban mobility policy: The impact in São Paulo transport system using system dynamics. *Transport Policy, 73*, 51–61.

Forrester, J. W. (1968). Industrial dynamics—after the first decade. *Management Science, 14*, 398–415.

Geels, F. W. (2004). From sectoral systems of innovation to socio-technical systems. *Research Policy, Elsevier BV, 33*(6–7), 897–920. https://doi.org/10.1016/j.respol.2004.01.015

Geels, F. W. (2011). The multi-level perspective on sustainability transitions: Responses to seven criticisms. *Environmental Innovation and Societal Transitions, Elsevier BV, 1*(1), 24–40. https://doi.org/10.1016/j.eist.2011.02.002

IEA. (2018). *Global EV outlook 2018: Towards cross-modal electrification*. IEA.

Kemp, R., & Loorbach, D. (2003). Governance for sustainability through transition management. In *Open meeting of human dimensions of global environmental change research community, Montreal, Canada* (S.l.: s.n., Vol. 20).

Loorbach, D., & Rotmans, J. (2010). *Transition management and strategic niche management*. [S.l.: s.n.]. Dutch Research Institute for Transitions.

Markard, J., Bento, N., Kittner, N., & Nuñez-Jimenez, A. (2020). Destined for decline? Examining nuclear energy from a technological innovation systems perspective. *Energy Research & Social Science, Elsevier BV, 67*, 101512. https://doi.org/10.1016/j.erss.2020.101512

Markard, J., Raven, R., & Truffer, B. (2012). Sustainability transitions: An emerging field of research and its prospects. *Research Policy, Elsevier BV, 41*(6), 955–967. https://doi.org/10.1016/j.respol.2012.02.013

Miller, J., & Façanha, C. (2016). *Cost-benefit analysis of Brazil's heavy-duty emission standards (P-8)*. International Council on Clean Transportation.

MMA. (2011). *1º Inventário Nacional de Emissões Atmosféricas por Veículos Automotores Rodoviários*. In Ambiente, M. D. M. (Ed.).

MMA. (2013). *Inventário Nacional de Emissões Atmosféricas por Veículos Automotores Rodoviários*.

NEB. (2018). *National energy balance, 2018: Base year 2017*.

Nilsson, M., & Nykvist, B. (2016). Governing the electric vehicle transition—Near term interventions to support a green energy economy. *Applied Energy, Elsevier BV, 179*, 1360–1371. https://doi.org/10.1016/j.apenergy.2016.03.056

Olsson, O., Grauers, A., & Pettersson, S. (2016). Method to analyze the cost-effectiveness of different electric bus systems. *29th World Electric Vehicle Symposium and Exhibition, EVS*.

ONU. (2019). *17 Objetivos para transformar nosso mundo* [Online]. Available: https://nacoesunidas.org/pos2015/agenda2030/. Accessed September 2019.

Raymundo, H., & Reis, J. G. M. (2015). Renovação da Frota de Ônibus Urbano: Redução de Consumo de Energia e de Impactos Ambientais. *5ª Academic International Workshop Advances in Cleaner Production*.

Roitman, T., & Da Silva, T. B. (2018). Concorrência interenergética e intermodal no setor de transportes: possibilidades para o Brasil. *Boletim de Conjuntura*, 15–23.

Rotmans, J., Kemp, R., & Van Asselt, M. (2001). More evolution than revolution: Transition management in public policy. *Foresight, 3*(1), 15–31.

Sayyadi, R., & Awasthi, A. (2017). A system dynamics-based simulation model to evaluate regulatory policies for sustainable transportation planning. *International Journal of Modelling Simulation, 37*, 25–35.

SCANIA. (2018). *SCANIA faz a maior venda de ônibus urbanos de sua história* [Online]. Available: https://www.scania.com/br/pt/home/experience-sca nia/news-and-events/News/archive/2018/11/default-press-release2.html. Accessed September 2019.

Segantin, C. C. (2019). *Barreiras e facilitadores para a implantação de ônibus elétrico no sistema de transporte público de São Paulo. Dissertação de Mestrado.* Universidade Nove de Julho.

Shepherd, S. (2014). A review of system dynamics models applied in transportation. *Transportmetrica B: Transport Dynamics, 2,* 83–105.

SIAMIG. (2018). *Biometano e a redução das emissões do transporte urbano* [Online]. Available: http://www.siamig.com.br/artigos/biometano-e-a-red ucao-das-emissoes-do-transporte-urbano. Accessed September 2019.

Slowik, P., Araujo, C., Dallmann, T., & Façanha, C. (2018). *Avaliação Internacional de Políticas Públicas para Eletromobilidade em Frotas Urbanas.* PROMOB-E.

SPTrans. (2019). *Valores das Tarifas Vigentes a partir de 07/01/2019* [Online]. SPTrans. Available: https://www.prefeitura.sp.gov.br/cidade/secretarias/tra nsportes/institucional/sptrans/acesso_a_informacao/index.php?p=227887. Accessed September 2019.

Tukker, A., & Butter, M. (2007). Governance of sustainable transitions: About the 4(0) ways to change the world. *Journal of Cleaner Production, Elsevier BV, 15*(1), 94–103. https://doi.org/10.1016/j.jclepro.2005.08.016

Vaz, C. R., & Maldonado, M. U. (2016). *O que é a dinâmica de sistemas? Reflexões sobre sua evolução e sobre as oportunidades de aplicação na Gestão de Operações.* SIMPOI.

Weber, K. M. (2003). Transforming large socio-technical systems towards sustainability: On the role of users and future visions for the uptake of city logistics and combined heat and power generation. *Innovation: The European Journal of Social Science Research, Informa UK Limited, 16*(2), 155–175. https://doi.org/10.1080/13511610304522

Wen, L., & Bai, L. (2017). System dynamics modeling and policy simulation for urban traffic: A case study in Beijing. *Environmental Modeling Assessment, 22,* 363–378.

INDEX